Mathematics Study Resources

Volume 9

Series Editors

Kolja Knauer, Departament de Matemàtiques Informàtic, Universitat de Barcelona, Barcelona, Barcelona, Spain
Elijah Liflyand, Department of Mathematics, Bar-Ilan University, Ramat-Gan, Israel

This series comprises direct translations of successful foreign language titles, especially from the German language.

Powered by advances in automated translation, these books draw on global teaching excellence to provide students and lecturers with diverse materials for teaching and study.

Stefan Schäffler

Mathematics
of Information

Theory and Applications
of Shannon-Wiener Information

 Springer

Stefan Schäffler
Fakultät für Elektro- und
Informationstechnik, Mathematik und
Operations Research
Universität der Bundeswehr München
Neubiberg, Germany

ISSN 2731-3824   ISSN 2731-3832   (electronic)
Mathematics Study Resources
ISBN 978-3-662-69101-4   ISBN 978-3-662-69102-1   (eBook)
https://doi.org/10.1007/978-3-662-69102-1

to my granddaughter **Emma Dorothea**

Introduction

> Information theory is a branch of the mathematical theory of probability and mathematical statistics.
> Solomon Kullback in [Kull97]

The term **information** is one of the key concepts of our time; sociologists therefore speak of the information age, among other things, when they describe the present. If one asks about the scientific disciplines that deal with information, most people probably think of computer science, telecommunications, and communication sciences in general (for example, as sub-areas of electrical engineering, psychology, and sociology); few think of mathematics. For this reason, information theory is no longer perceived as a subfield of mathematics. This is all the more surprising when one considers that it was mathematicians who did the pioneering work of a scientific theory of information—embedded in stochastics.

In a mathematical-scientific context, the term information probably first appeared in 1925 in a work by RONALD AYLMER FISHER (1890–1962) titled "Theory of statistical estimation" (see [Fi25]); in principle, it asks about the amount of information one obtains about unknown distribution parameters through realizations of the underlying random experiment; we will address a related question when we discuss sufficient statistics. In this book, an approach to the concept of information alternative to FISHER-information is chosen. The first publication in this context can be considered the article "A Mathematical Theory of Communication", the first part of which was published by CLAUDE ELWOOD SHANNON (1916–2001) in July 1948 in the *Bell Systems Technical Journal* (see [ShWe63]). The aim of this approach is clear from the title of the article: The concept of information serves as a building block for a theory of communication to be developed. In the same year, NORBERT WIENER (1894–1964) independently chose an analogous approach for continuous distributions by introducing differential entropy (see [Wie61]), which had already been used by LUDWIG BOLTZMANN (1844–1906) in a thermodynamic context in 1866; however, WIENER explicitly used the term information in this context, unlike BOLTZMANN and SHANNON. This book thus focuses on the Shannon-Wiener approach to the mathematical concept of information. As NORBERT WIENER already noted in his book on cybernetics ([Wie61]), this approach can replace the concept of information by FISHER (the corresponding tool is the "de Bruijn identity", see for example [Joh04]). Although both SHANNON and WIENER primarily had communication theory in mind when designing the mathematical "measurement" of an amount of information, the parallel to thermodynamics already shows that the concept of information to be examined in the following is much more broadly applicable.

The first part of this book is dedicated to the basics. It is essential to assign an amount of information to a probability after a mutual delimitation of the two terms message and information.

In part 2 countable systems are examined, i.e., probability spaces with at most countably many results. The average amount of information of these systems leads to the fundamental definition of Shannon entropy and its application as a suitable measure for the quality of a coding; Huffman coding is presented as an optimal coding. A very important example of countable systems is provided by statistical physics—more precisely thermodynamics. In this context, Shannon entropy allows for an information-theoretical interpretation of thermodynamic entropy and in particular of the second law of thermodynamics for closed systems (Chap. 4). With the introduction of conditional probabilities in Chap. 5 two classic applications of Shannon entropy in mathematical statistics (sufficiency of estimation functions) and in communication technology (transinformation) are presented. A book about the mathematics of information would be incomplete without a look at quantum information and quantum algorithms; the corresponding sixth chapter serves as an introduction to this topic.

Part 3 begins the analysis of general systems, that is, of probability spaces with generally more than countably many results; this leads to the concept of differential entropy introduced by NORBERT WIENER in the context of information theory. The necessary measure-theoretical and integration-theoretical prerequisites are developed at the point where they are needed. In addition to applications from communications engineering and mathematical statistics, the concept of information is also considered in the analysis of dynamic systems.

This book is written for mathematicians and/or computer scientists with prior knowledge at the bachelor's level. On the one hand, it serves to illustrate how the concept of information is anchored in mathematics, and on the other hand, it is intended to present a selection of applications (especially outside of communication technology); as it is designed as a mathematics book, great emphasis is naturally placed on exact proof. Therefore, the content of this book will not coincide with the contents that engineers associate with the term information theory (there is an overwhelming amount of literature on this; [CovTho91] is mentioned as a classic). To avoid misunderstandings, the term information theory was therefore avoided in the title of this book. Nevertheless, the author remains convinced that even engineers are able to benefit from reading this book if they engage with the language and presentation style of mathematics, which unfortunately is not a given in Germany.

But mathematicians will also miss certain topics in this book; I am thinking in particular of ergodic theory . This topic has long since developed into its own special discipline; in Sect. 7.3 and 9.4 we will at least discuss the concept of the ergodic probability measure. A standard work on the subject of ergodic theory and information is still [Bill65]. For a mathematically sound introduction to coding theory, which is also only considered here in passing, refer to [Ash65] and [HeiQua95]. There is a very interesting connection between Shannon entropy and the Hausdorff dimension of a set; here refer to [Bill65] and [PötSob80].

My esteemed colleague and employee, Mr. Dr. R. von CHOSSY has critically worked through the manuscript in the last weeks of his service, brought in many suggestions for improvement and valuable proof ideas (for example, for Theorem 8.1) and was thus, as always, an invaluable help to me. I would like to express my special thanks to him at this point—especially for his excellent work over the past 14 years.

Contents

Symbols

$\mathcal{P}(\Omega)$	Power set of Ω
$\mathbb{S}$	Entropy
$\mathbb{P}_X$	Image measure of X
$\mathbb{P}^B$	conditional probability
$\mathbb{T}$	Transinformation
C	Channel capacity
R	Code rate
$\mathcal{H}$	Hilbert space
$\langle \cdot, \cdot \rangle_{\mathcal{H}}$	scalar product in $\mathcal{H}$
$\mathcal{S}_{\mathcal{H}}$	Sphere of $\mathcal{H}$
$\mathcal{H}^{\otimes n}$	n-fold tensor product of $\mathcal{H}$
$\sigma(X)$	σ-algebra generated by X
$\| \cdot \|_{\infty}$	maximum norm
$\mathcal{Z}$	Cylinder set
$\oplus$	binary addition
$(\Omega, \mathcal{S})$	measurable space
$(\Omega, \mathcal{S}, \mathbb{P})$	probability space
$\mathfrak{F}$	Frobenius-Perron Operator
$\mathbb{E}(X|\mathcal{C})$	conditional expectation

List of Figures

List of Tables

Part I
Basics

Message and Information

1

1.1   Starting Point Transmitter: Message

The two terms **message** and **information** are components of our everyday language that are not always precisely distinguished. However, within the context of engineering and mathematics, it is essential to clearly differentiate these two terms. A message is initially something that always originates from a transmitter and is put into a specific physical form. This physical form depends on the way the corresponding message is to be transmitted from the transmitter to the intended recipients. For example, among the Native American tribes, messages were transmitted using specific smoke signals. Since about 1817, messages in shipping have been exchanged, among other things, through flag signals. The representation of a message depending on the intended type of transmission thus plays an important role in communication technology.

A second important factor is the security of the transmission. The representation of a message often has to be chosen in such a way that, on the one hand, unauthorized persons cannot understand the content of the message, even though they have intercepted the message itself (encryption, cryptography), and on the other hand, in the event of unavoidable transmission errors, the intended recipients should still be able to reconstruct the correct content of the message (channel coding). A third factor is economy: a message should be represented in such a way that as few resources as possible are needed for transmission. In the case of radio transmission, the resources would be, for example, time (signal duration) and bandwidth (required frequencies). Let's consider, for example, the most common type of communication, sending an SMS. First, a message is formulated as text in (German) language. Let's choose as an example a message that is certainly sent millions of times a day among teenagers in Germany:

MATHEMATIK IST SCHÖN

© The Author(s), under exclusive license to Springer-Verlag GmbH, DE, part of
Springer Nature 2024
S. Schäffler, *Mathematics of Information*, Mathematics Study Resources 9,
https://doi.org/10.1007/978-3-662-69102-1_1

(engl.: Mathematics is beautiful). This message cannot be transmitted by radio as it is; moreover, it is not yet protected. In order to be able to protect a message, it must be put into an **algebraic** form; that is, it must be transformed in such a way that calculations can be made with the symbols that form the message. One could, for example, come up with the idea of assigning each letter of the German alphabet including umlauts and spaces a finite sequence of bits $b_i \in \{0, 1\}$, because the set $\{0, 1\}$ has an easy-to-implement algebraic structure through

$$0 \oplus 0 = 1 \oplus 1 = 0 \quad \text{and} \quad 0 \oplus 1 = 1 \oplus 0 = 1.$$

Such a chosen finite sequence is referred to as the code of the corresponding character from the alphabet. In this process, however, it is important to ensure, on the one hand, that the message transformed into a sequence of bits can be uniquely reconstructed, and on the other hand, for reasons of economy, to use as few bits as possible. The above Table 1.1 now provides a coding of the characters of our alphabet. Based on a statistical evaluation of various texts in German, a probability of occurrence W was determined for each character from the alphabet. The argument for economy is now taken into account by using fewer bits for the coding of a character, the greater the probability of occurrence; otherwise, the at first glance unexpected coding (for example, that each character is coded with at least three bits) is due to the unique reconstructability of the message.

Table 1.1 Binary coding; Source: [Kom]

No.	Character	W	Code	No.	Character	W	Code
1	"Space"	0.15149	000	16	O	0.01772	111001
2	E	0.14700	001	17	B	0.01597	111010
3	N	0.08835	010	18	Z	0.01423	111011
4	R	0.06858	0110	19	W	0.01420	111100
5	I	0.06377	0111	20	F	0.01360	111101
6	S	0.05388	1000	21	K	0.00956	1111100
7	T	0.04731	1001	22	V	0.00735	1111101
8	D	0.04385	1010	23	Ü	0.00580	11111100
9	H	0.04355	10110	24	P	0.00499	11111101
10	A	0.04331	10111	25	Ä	0.00491	11111110
11	U	0.03188	11000	26	Ö	0.00255	111111110
12	L	0.02931	11001	27	J	0.00165	1111111110
13	C	0.02673	11010	28	Y	0.00017	11111111110
14	G	0.02667	11011	29	Q	0.00015	111111111110
15	M	0.02134	111000	30	X	0.00013	111111111111

The average number L of bits per character results in this encoding

$$
\begin{aligned}
L = {} & 0.15149 \cdot 3 + 0.14700 \cdot 3 + 0.08835 \cdot 3 \\
& + 0.06858 \cdot 4 + 0.06377 \cdot 4 + 0.05388 \cdot 4 \\
& + 0.04731 \cdot 4 + 0.04385 \cdot 4 + 0.04355 \cdot 5 \\
& + 0.04331 \cdot 5 + 0.03188 \cdot 5 + 0.02931 \cdot 5 \\
& + 0.02673 \cdot 5 + 0.02667 \cdot 5 + 0.02134 \cdot 6 \\
& + 0.01772 \cdot 6 + 0.01597 \cdot 6 + 0.01423 \cdot 6 \\
& + 0.01420 \cdot 6 + 0.01360 \cdot 6 + 0.00956 \cdot 7 \\
& + 0.00735 \cdot 7 + 0.00580 \cdot 8 + 0.00499 \cdot 8 \\
& + 0.00491 \cdot 8 + 0.00255 \cdot 9 + 0.00165 \cdot 10 \\
& + 0.00017 \cdot 11 + 0.00015 \cdot 12 + 0.00013 \cdot 12 \\
= {} & 4.14834 \text{ Bits.}
\end{aligned}
$$

We will come back to this value later.
Now, if we use this encoding, our original message

MATHEMATIK IST SCHÖN

becomes the bit sequence

111000101111001101100011110010
111100101111111100000011110010
010001000110101011011111110010

consisting of 93 bits. To protect this message from unauthorized reading (we assume that everyone knows that Table 1.1 is used), we choose randomly generated 93 bits, for example

010011100111110110110111011110
000100101111100000111011010110.
001011000111100101101000010100

This sequence of bits is referred to as a key and should only be known to the sender and all recipients who are authorized to read the sent message. We now add these two sequences of bits bit by bit (i.e., first bit with first bit, second bit with second bit, etc.) using the binary addition introduced above $\oplus$. The result we get is:

101011001000111011010101001011100
111000000000011100111000100110.
011010000001001110110111011110

Let's assume that this message was transmitted without errors. All recipients who are authorized to read this message know the used key and now add this key bit by bit to the received message. The result is the already known bit sequence

$$111000101111001101100011110010$$
$$111100101111111100000011110010,$$
$$010001000110101011011111110010$$

which, using Table 1.1, leads to

MATHEMATIK IST SCHÖN.

An unauthorized recipient who has intercepted the message does not know the key and, using Table 1.1 applied to the intercepted message

$$101011001000111011010101001011100$$
$$111000000000011100111000100110,$$
$$011010000001001110110111101111110$$

arrives at the useless result

DL ZNTIEU IEUTSC EEGIA,

where the last two bits no longer make sense.

This type of encryption is perfectly secure, but impractical in practice, as it uses keys that are as long as the message itself. These keys must be known to the transmitter and the authorized recipients and must above all remain secret (i.e., they must not fall into the hands of unauthorized persons through espionage); therefore, cryptography today uses algebraic, geometric, and number-theoretical methods of mathematics, which, however, are not the subject of this book (see, for example, [For13] and [Buch10]).

Since bits cannot be transmitted wirelessly, the technique of modulation is required. The message consisting of 93 bits $b_1, \ldots, b_{93}$ is transformed into a signal s, which can now be transmitted in the form of an electromagnetic wave; a very simple type of modulation is given by the following signal, where $N \in \mathbb{N}$ and the frequency f is chosen:

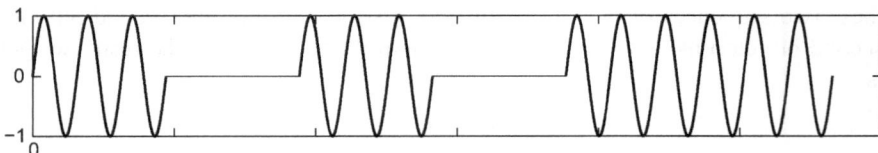

Fig. 1.1 Amplitude modulation

$$s : \left[0, \frac{N}{f} \cdot 93\right) \to \mathbb{R}, \quad t \mapsto \sum_{i=1}^{93} I_{[0,1)}\left(\frac{tf}{N} - (i-1)\right) b_i \sin(2\pi ft)$$

with

$$I_{[0,1)} : \mathbb{R} \to \{0, 1\}, \quad x \mapsto \begin{cases} 1 & \text{if } 0 \le x < 1 \\ 0 & \text{if } x < 0 \text{ or } x \ge 1 \end{cases}.$$

For each bit, N periods of the sine function with frequency f are reserved in sequence. If a bit has the value one, the amplitude is chosen to be one in the corresponding interval. If a bit has the value zero, the amplitude is also chosen to be zero in the corresponding interval. For the first six bits 101011, Fig. 1.1 shows the corresponding part of the signal s with $N = 3$.

Of course, this is a very simple type of modulation and not state of the art (see [OhmLü10]); it is only about the basic procedure here. As with any transmission, there can also be disturbances in radio transmission, so depending on the size of the disturbance, a signal like in Fig. 1.2 (for the first six bits) will arrive at the receiver instead of the signal s.

Therefore, it is not surprising that errors can occur during the back transformation from the signal to the bits (demodulation). For example, if only the third bit flips, only the third bit is wrong after adding the key and the bit sequence

$$11000010111100110110001111100010$$
$$11110010111111110000001111100010$$
$$01000100011010101101111111110010$$

is present in the receiver, which leads to the text

UNWG WNWAK IST SCHÖN

by using Table 1.1. Therefore, it is necessary to take protective measures in the transmitter to be able to detect and correct flipped bits in the receiver. The protection now consists in transmitting the bits not only once, but several times. However, a simple repetition is not efficient. Therefore, we consider as an example a channel coding that protects every four consecutive bits with three bits. So, each block of four

$$b_1 b_2 b_3 b_4 \quad b_5 b_6 b_7 b_8 \quad \ldots \quad b_{89} b_{90} b_{91} b_{92}$$

is supplemented as follows by three bits:

$$b_1 b_2 b_3 b_4 \quad \text{is becoming} \quad b_1 b_2 b_3 b_4 \underbrace{(b_2 \oplus b_3 \oplus b_4)}_{c_1} \underbrace{(b_1 \oplus b_3 \oplus b_4)}_{c_2} \underbrace{(b_1 \oplus b_2 \oplus b_4)}_{c_3}.$$

The bit c_1 is therefore obtained by the sum of the second, third and fourth bits in the block of four, the bit c_2 by the sum of the first, third and fourth bits in the

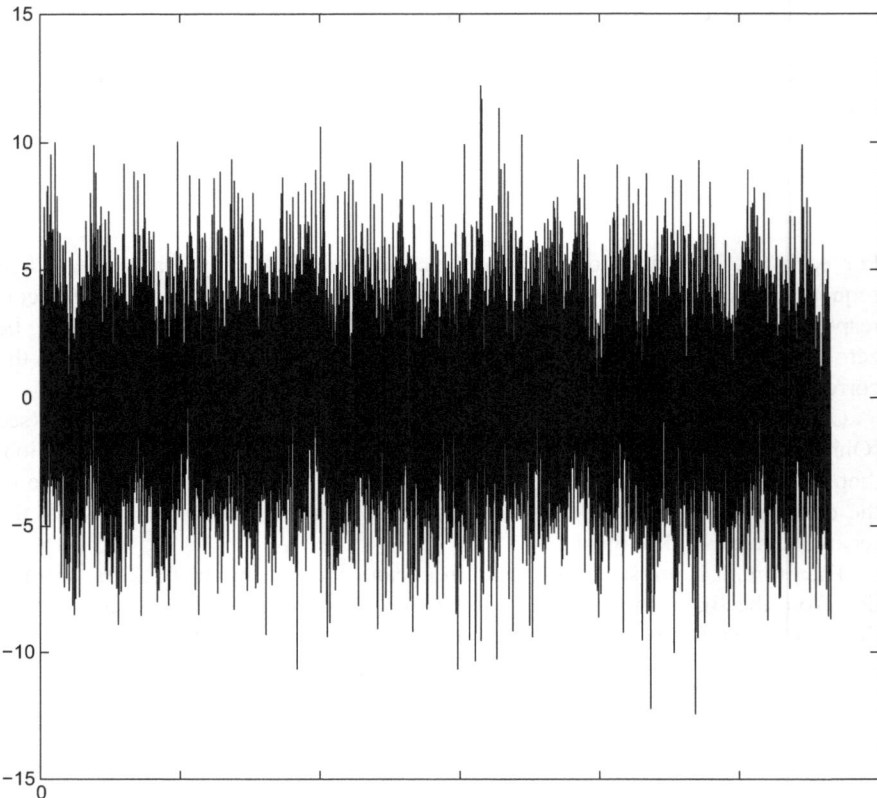

Fig. 1.2 Amplitude modulation with disturbance

block of four, and the bit c_3 through the sum of the first, second, and fourth bits in the four-bit block. The fifth four-bit block thus becomes

$$b_{17}b_{18}b_{19}b_{20} \;\rightarrow\; b_{17}b_{18}b_{19}b_{20} \underbrace{(b_{18} \oplus b_{19} \oplus b_{20})}_{c_{13}} \underbrace{(b_{17} \oplus b_{19} \oplus b_{20})}_{c_{14}} \underbrace{(b_{17} \oplus b_{18} \oplus b_{20})}_{c_{15}}.$$

In a seven-bit block such as $b_1b_2b_3b_4c_1c_2c_3$, b_1 appears three times:

(i) directly in the first position,
(ii) indirectly in $c_2 = b_1 \oplus b_3 \oplus b_4$,
(iii) indirectly in $c_3 = b_1 \oplus b_2 \oplus b_4$.

The bit b_2 appears three times:

(i) directly in the second position,
(ii) indirectly in $c_1 = b_2 \oplus b_3 \oplus b_4$,
(iii) indirectly in $c_3 = b_1 \oplus b_2 \oplus b_4$.

The bit b_3 appears three times:

(i)   directly in the third position,
(ii) indirectly in $c_1 = b_2 \oplus b_3 \oplus b_4$,
(iii) indirectly in $c_2 = b_1 \oplus b_3 \oplus b_4$.

The bit b_4 appears four times:

(i)   directly in the fourth position,
(ii) indirectly in $c_1 = b_2 \oplus b_3 \oplus b_4$,
(iii) indirectly in $c_2 = b_1 \oplus b_3 \oplus b_4$,
(iv) indirectly in $c_3 = b_1 \oplus b_2 \oplus b_4$.

Therefore, instead of the 93 bits

$$1010110010001110110101001011100$$
$$11100000000001110011100010011 00,$$
$$01101000000100111011011110111110$$

the 162 bits (**bold** are the newly added bits)

$$10101\mathbf{011}10\mathbf{011}01000011\mathbf{111}1\mathbf{0000}110\mathbf{1}0010100\mathbf{1}011011\mathbf{010}$$
$$1001\mathbf{1001}100\mathbf{1}1000000000000000\mathbf{0}1110000011111000001111$$
$$0011\mathbf{001}00011111010101000000000100101111\mathbf{0000}1101\mathbf{001}$$
$$111\mathbf{0000}11111\mathbf{110}$$

must be transmitted, where one could, for example, decide to leave the 93rd bit unprotected. This artificially added redundancy allows correcting a single error that occurs in a seven-bit block:

If, for example, the third bit flips in the first block of seven, you get

$$1000\mathbf{101} \quad \text{instead of} \quad 1010\mathbf{101}.$$

The combination 1000**101** is not provided; so something must have happened during the transmission. Now let's assume that exactly one bit has flipped, so there are seven variants:

$$
\begin{array}{ll}
0000\mathbf{101} & \text{not provided,} \\
1100\mathbf{101} & \text{not provided,} \\
1010\mathbf{101} & \text{possible,} \\
1001\mathbf{101} & \text{not provided,} \\
1000\mathbf{001} & \text{not provided,} \\
1000\mathbf{111} & \text{not provided,} \\
1000\mathbf{100} & \text{not provided.}
\end{array}
$$

Thus, the third bit can be detected as wrong. The procedure just outlined is a simple example (a so-called Hamming code) for a channel coding, to add redundancy in the transmitter using algebraic operations, so that errors can be corrected in the receiver. Of course, this idea costs resources, as instead of 93 now 162 bits are transmitted. To find channel codings that require few resources on the one hand and have good correction capabilities on the other hand, deep geometric and algebraic knowledge is needed (see for example [Frie96]).

1.2   Endpoint Receiver: Information

In physics, there are quantities (such as **time** or **mass**) that are considered basic building blocks of nature description and therefore are not defined. According to BLAISE PASCAL (1623–1662), this is also not necessary, as every human being has the same—innate—understanding of these basic quantities and a definition, which would always have to rely on already given quantities, would even be harmful (see [Pas38])[1]. However, one can name essential properties of these quantities and one can also measure these basic quantities.

The term **information** behaves analogously; a formal definition is also not sensible, but one can specify properties and we will measure the **amount of information** (or synonymously the **information content**) in a mathematical sense. While a message is created in a transmitter and then transmitted to a receiver, information always arises in a receiver through the receipt of a message—and that is when the content of the received message is associated with a surprise effect for the receiver (the transmitter is also part of the content of a message). The greater the surprise that the content of a message triggers in the receiver, the more information the receiver has received from this message (so the greater the transmitted amount of information). Let's return to the SMS about the beauty of mathematics as an example. If I, as a mathematician, were to send this SMS to my children, this message would contain no information for the respective receivers, as they know that the SMS comes from me and as they know my opinion about mathematics exactly. Conversely, if one of my children were to send this SMS to me, the content of this message would have an infinitely large information content for me due to my knowledge of the transmitter, because the surprise would be just as great. The question now arises as to how one can quantify the surprise effect that a message triggers in a receiver. The representation of the message as a sequence of bits in connection with the algebraic structure—given by the operation $\oplus$—has led to the fact that one can fall back on the valuable results of algebra and number theory for the given goal. It therefore makes sense to choose a procedure for quantifying

---

[1] In the middle of the 17th century, other basic quantities were discussed than today. However, **time** was already mentioned by PASCAL in this context.

the surprise effect that a message triggers in a receiver that allows the application of an advanced mathematical theory. For this reason, it was decided to consider the probability with which a receiver expects the content of a message (including transmitter). The smaller this probability is, the greater is the surprise upon receipt of the message (and thus the greater the information content of the message for this receiver). This makes the results of stochastics usable for a "mathematics of information" to be developed.

Since we only consider the probability with which a receiver expected a received message for determining its information content, we can define the term **message** very broadly. A message is any event that occurs with a certain probability. In the following chapter, we will therefore establish a connection between a probability, i.e., a real number from the interval [0, 1], and the associated amount of information.

Information and Chance

2

2.1 Probability and Amount of Information

Next, we are looking for a function I defined on the interval $[0, 1]$, which assigns an amount of information $I(p)$ to each probability $p \in [0, 1]$; certain properties are required of this function I:

(I1) The function $I : [0, 1] \to [0, \infty]$ should be continuous on the open interval $(0, 1)$.

This requirement probably needs no explanation. No one will seriously demand or allow discontinuities.

(I2) $I\left(\frac{1}{2}\right) = 1$.

One can only measure if one has established a unit (like the original "meter" as the unit of length measurement in Paris). This requirement now sets the information quantity one as the unit for the probability $\frac{1}{2}$.

(I3) $I(pq) = I(p) + I(q)$ for all $p, q \in (0, 1)$.

If an event A occurs with probability p and an event B occurs with probability q, these events are considered stochastically independent if the joint occurrence of A and B occurs with probability pq. In this case, the occurrence of A does not influence the probability of the occurrence of B and vice versa. It is therefore sensible to define the amount of information that includes the joint occurrence of A and B as the sum of the individual amounts of information (from A and from B).

(I4) $I(0) = \lim\limits_{\substack{p \to 0 \\ p \in (0,1)}} I(p), \ I(1) = \lim\limits_{\substack{p \to 1 \\ p \in (0,1)}} I(p)$

This fourth requirement is again an argument for continuity.

© The Author(s), under exclusive license to Springer-Verlag GmbH, DE, part of Springer Nature 2024
S. Schäffler, *Mathematics of Information*, Mathematics Study Resources 9,
https://doi.org/10.1007/978-3-662-69102-1_2

Now, in a first result, it should be shown that the function I on the interval $(0, 1)$ is uniquely determined by the first three properties.

▶ **Uniqueness of the Function I**

There is exactly one function

$$h : (0, 1) \to (0, \infty)$$

with:

(i) h is continuous.
(ii) $h(\frac{1}{2}) = 1$.
(iii) $h(pq) = h(p) + h(q)$ for all $p, q \in (0, 1)$.

This function is the inverse function to

$$f : (0, \infty) \to (0, 1), \quad x \mapsto 2^{-x}$$

and thus the negative binary logarithm on the interval $(0, 1)$ (denoted by: $- \operatorname{ld}_{(0,1)}$). It holds:

$$\lim_{\substack{x \to 0 \\ x > 0}} x \cdot (-\operatorname{ld}_{(0,1)}(x)) = 0.$$

Let $n, m \in \mathbb{N}$, then for a function

$$h : (0, 1) \to (0, \infty)$$

with the properties (i)–(iii) holds:

$$h(2^{-n}) = h\left(\left(\frac{1}{2}\right)^n\right) = n \cdot h\left(\frac{1}{2}\right) = n.$$

Furthermore, we obtain from

$$n = h\left(\left(\frac{1}{2}\right)^n\right) = h\left(\left(\left(\frac{1}{2}\right)^{\frac{n}{m}}\right)^m\right) = mh\left(\left(\frac{1}{2}\right)^{\frac{n}{m}}\right)$$

the equation

$$h(2^{-\frac{n}{m}}) = h\left(\left(\frac{1}{2}\right)^{\frac{n}{m}}\right) = \frac{n}{m}.$$

Now let $y \in (0, 1)$, then there is a unique $x \in (0, \infty)$ with $y = 2^{-x}$. Since $\mathbb{Q}$ is dense in $\mathbb{R}$, there are two sequences $\{m_i\}_{i \in \mathbb{N}}$ and $\{n_i\}_{i \in \mathbb{N}}$ of natural numbers with

$$\lim_{i \to \infty} \frac{n_i}{m_i} = x.$$

From the continuity of f and h it follows:

$$h(2^{-x}) = h\left(2^{-\lim\limits_{i\to\infty}\frac{n_i}{m_i}}\right) = h\left(\lim_{i\to\infty} 2^{-\frac{n_i}{m_i}}\right)$$

$$= \lim_{i\to\infty} h\left(2^{-\frac{n_i}{m_i}}\right) = \lim_{i\to\infty} h\left(\left(\frac{1}{2}\right)^{\frac{n_i}{m_i}}\right)$$

$$= \lim_{i\to\infty} \frac{n_i}{m_i} = x.$$

So it holds: $h = -\operatorname{ld}_{(0,1)}$. For $x>0$ we know

$$e^x = \sum_{k=0}^{\infty} \frac{x^k}{k!} = 1 + x + \frac{x^2}{2} + \sum_{k=3}^{\infty} \frac{x^k}{k!} > \frac{x^2}{2}.$$

Thus we obtain:

$$0 \le \lim_{\substack{x\to 0 \\ x>0}} x \cdot (-\operatorname{ld}_{(0,1)}(x)) = \lim_{y\to\infty} \frac{1}{y\ln(2)} \cdot \left(-\ln_{(0,1)}\left(\frac{1}{y}\right)\right) = \lim_{y\to\infty} \frac{\ln(y)}{y\ln(2)}$$

$$= \lim_{y\to\infty} \frac{\ln(y)}{e^{\ln(y)}\ln(2)} = \lim_{z\to\infty} \frac{z}{e^z\ln(2)} \le \lim_{z\to\infty} \frac{z}{\frac{z^2}{2}\ln(2)}$$

$$= 0.$$

q.e.d

From this result, it follows that our sought function I is defined on the interval $(0, 1)$ by the function $-\operatorname{ld}_{(0,1)}$. Since

$$\lim_{\substack{p\to 1 \\ p\in(0,1)}} (-\operatorname{ld}_{(0,1)}(p)) = -\operatorname{ld}(1) = 0 \quad \text{and}$$

$$\lim_{\substack{p\to 0 \\ p\in(0,1)}} (-\operatorname{ld}_{(0,1)}(p)) = \lim_{\substack{p\to 0 \\ p\in(0,1)}} (-\operatorname{ld}(p)) = \infty,$$

follows (see Fig. 2.1):

$$I(0) = \infty \quad \text{and} \quad I(1) = 0.$$

With the usual definition in measure theory

$$\infty + a = a + \infty = \infty \quad \text{for all} \quad a \in \mathbb{R} \cup \{\infty\}$$

even

$$I(pq) = I(p) + I(q) \quad \text{for all} \quad p, q \in [0, 1].$$

applies.

The amount of information also has a unit; it is measured in **bit**. This choice is obvious when considering the following special cases, where the index "b" means that the binary system is used as a basis:

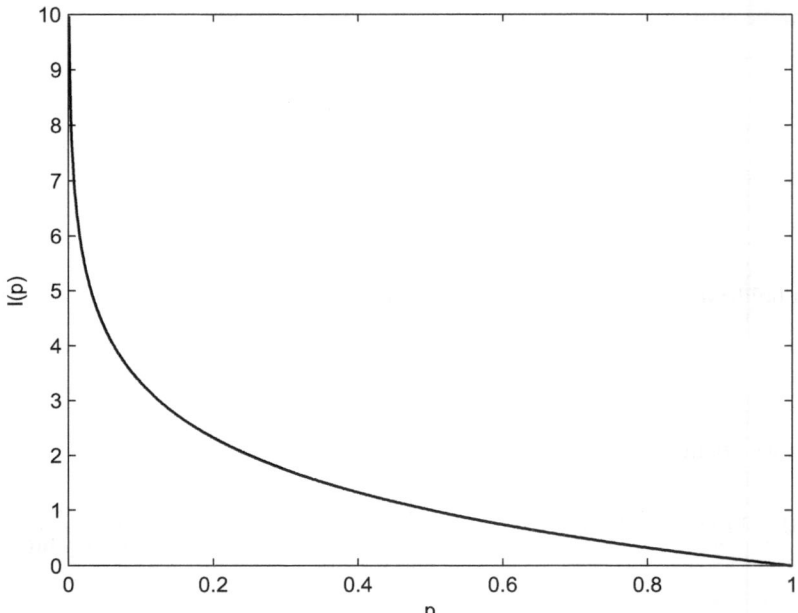

Fig. 2.1 The function I

$$I\left(\frac{1}{2}\right) = I(0.1_b) = 1 \text{ bit,}$$

$$I\left(\left(\frac{1}{2}\right)^k\right) = I(0.\underbrace{0\ldots01_b}_{k \text{ digits}}) = k \text{ bit,}$$

$$3 \text{ bit} = I\left(\frac{1}{8}\right) = I(0.001_b)$$

$$\leq I(0.\overline{0001}_b) = I\left(\frac{1}{15}\right) = 3.9069\ldots \text{ bit}$$

$$< I(0.0001_b) = 4 \text{ bit.}$$

The amount of information of a number $p \in (0, 1]$ can therefore be estimated with

$$\lfloor p \rfloor := \min\left\{k \in \mathbb{N}_0; \ \left(\frac{1}{2}\right)^k \geq p\right\}$$

by

$$\lfloor p \rfloor \leq I(p) < \lfloor p \rfloor + 1.$$

Since the functions $- \mathrm{ld}_{(0,1)}$ and $- \mathrm{ld}$ do not differ on the interval $(0,1)$, we will only use the function $- \mathrm{ld}$ or ld in the following. If we had set up the requirement $I\left(\frac{1}{\zeta}\right) = 1$ instead of $I\left(\frac{1}{2}\right) = 1$ for $\zeta > 1$ in requirement [I2], we would have obtained the logarithm to the base ζ instead of the binary logarithm.

To get an idea of the concept of **information quantity** given by the function I, let's consider the following

Example 2.2 On May 31, 2010, two people, A and B, receive a message from a third person—let's call them C—that today German Federal President Horst Köhler has resigned. Person A already knew this, while person B knew nothing and thought the resignation of a Federal President was impossible. Thus, the same message contains completely different amounts of information for persons A and B. For person A, the probability p_A that Horst Köhler resigns, at the moment she receives the message from person C, is equal to one, because she already knew the content of the message. Therefore, the message was not associated with any information:

$$I(p_A) = I(1) = - \mathrm{ld}(1) = 0.$$

For person B, the surprise was infinitely great, as she considered this resignation impossible ($p_B = 0$):

$$I(p_B) = I(0) = \lim_{\substack{x \to 0 \\ x > 0}} - \mathrm{ld}(x) = \infty.$$

Person C had another message ready, namely that on the same day the Israeli army had boarded a ship convoy of the Free Gaza Movement. Both persons A and B expected this action with probability $q_A = q_B = 0.75$ as the state of Israel had already announced this procedure several times, but they did not know anything about it yet. Intuitively, one would set the total amount of information that person A has received from these two messages to

$$I(1) + I(0.75) = 0 - \mathrm{ld}(0.75) \approx 0.415 \text{ bit.}$$

This is because both events (resignation of the German Federal President and military action of Israel) do not influence each other. The probability of both events occurring is therefore equal to $p_A q_A$ for person A and $p_B q_B$ for Person B and it holds for person A due to (iii):

$$I(p_A q_A) = I(p_A) + I(q_A) = 0 - \mathrm{ld}(0.75) = - \mathrm{ld}(0.75) \approx 0.415 \text{ bit.}$$

What does the total amount of information for person B look like now? Due to $0 \cdot 0.75 = 0$ and due to the definition

$$\infty + a = a + \infty = \infty \quad \text{for all} \quad a \in \mathbb{R} \cup \{\infty\}$$

it holds:

$\triangleleft$
$$\infty = I(0) = I(p_B q_B) = I(0 \cdot 0.75) = I(0) + I(0.75) = \infty - \mathrm{ld}(0.75) = \infty.$$

Table 2.1 Information quantity per character

No.	Character	W	$-\mathrm{ld}(W)$ [bit]	No.	Character	W	$-\mathrm{ld}(W)$ [bit]
1	"Space"	0.15149	2.72271	16	O	0.01772	5.81848
2	E	0.14700	2.76611	17	B	0.01597	5.96849
3	N	0.08835	3.50063	18	Z	0.01423	6.13492
4	R	0.06858	3.86607	19	W	0.01420	6.13797
5	I	0.06377	3.97098	20	F	0.01360	6.20025
6	S	0.05388	4.21411	21	K	0.00956	6.70877
7	T	0.04731	4.40171	22	V	0.00735	7.08804
8	D	0.04385	4.51128	23	Ü	0.00580	7.42973
9	H	0.04355	4.52118	24	P	0.00499	7.64674
10	A	0.04331	4.52916	25	Ä	0.00491	7.67006
11	U	0.03188	4.97120	26	Ö	0.00255	8.61529
12	L	0.02931	5.09246	27	J	0.00165	9.24331
13	C	0.02673	5.22540	28	Y	0.00017	12.52218
14	G	0.02667	5.22864	29	Q	0.00015	12.70275
15	M	0.02134	5.55030	30	X	0.00013	12.90920

2.2 The Average Information Quantity of a Character

As summarized in Table 1.1, in a German text, each character has a certain probability of occurrence. If you now select any place in a German text and look at the character that is there, you get a certain amount of information by recognizing this character, which is documented in the following Table 2.1.

The average amount of information $\bar{I}$ per character now results in

$$
\begin{aligned}
\bar{I} = {}& 0.15149 \cdot 2.72271 + 0.14700 \cdot 2.76611 + 0.08835 \cdot 3.50063 \\
& + 0.06858 \cdot 3.86607 + 0.06377 \cdot 3.97098 + 0.05388 \cdot 4.21411 \\
& + 0.04731 \cdot 4.40171 + 0.04385 \cdot 4.51128 + 0.04355 \cdot 4.52118 \\
& + 0.04331 \cdot 4.52916 + 0.03188 \cdot 4.97120 + 0.02931 \cdot 5.09246 \\
& + 0.02673 \cdot 5.22540 + 0.02667 \cdot 5.22864 + 0.02134 \cdot 5.55030 \\
& + 0.01772 \cdot 5.81848 + 0.01597 \cdot 5.96849 + 0.01423 \cdot 6.13492 \\
& + 0.01420 \cdot 6.13797 + 0.01360 \cdot 6.20025 + 0.00956 \cdot 6.70877 \\
& + 0.00735 \cdot 7.08804 + 0.00580 \cdot 7.42973 + 0.00499 \cdot 7.64674 \\
& + 0.00491 \cdot 7.67006 + 0.00255 \cdot 8.61529 + 0.00165 \cdot 9.24331 \\
& + 0.00017 \cdot 12.52218 + 0.00015 \cdot 12.70275 + 0.00013 \cdot 12.90920 \\
= {}& 4.11461 \text{ bit.}
\end{aligned}
$$

If we compare this result with the average word length calculated from Table 1.1 $L = 4.14834$ Bits, it becomes apparent that the encoding in Table 1.1 can practically not be improved any further (see Sect. 3.3).

Part II
Countable Systems

The Entropy

3

3.1 Discrete Probability Spaces

In this chapter, we consider random experiments, that is, experiments where, on the one hand, we know exactly which results are possible, but on the other hand, we can generally predict a result not exactly, but only with a certain probability. In the last section, such an experiment was described by examining a specific location in a German text to see which character is there. We based this on an alphabet of 30 different characters; thus, for this experiment, there were 30 different results and the corresponding probabilities available. This example shows that the term **experiment** is very broadly defined and is not limited to scientific experiments. The reception of a message is also considered an experiment in this context, with the result of the experiment being the message itself. The special feature of the random experiments in this chapter is that the non-empty set of possible results, always denoted by Ω, may only contain a finite number or countably infinite number of elements. Thus, there is a subset $N \subseteq \mathbb{N}$ such that a bijection $N \to \Omega$ exists. Now, if for each $\omega \in \Omega$ the probability $\mathbb{P}(\{\omega\})$ is known that we obtain ω as the result of the random experiment, we can assign a probability to each subset $A \subseteq \Omega$ of Ω by

$$\mathbb{P}(A) := \sum_{\omega \in A} \mathbb{P}(\{\omega\})$$

that the result of the random experiment is in the set A. If we naturally demand $\mathbb{P}(\Omega) = 1$ and $\mathbb{P}(\emptyset) = 0$, we obtain a mapping $\mathbb{P}$ with the following properties:

(P1) $\mathbb{P} : \mathcal{P}(\Omega) \to [0, 1]$, where $\mathcal{P}(\Omega)$ denotes the power set of Ω.
(P2) $\mathbb{P}(\emptyset) = 0, \mathbb{P}(\Omega) = 1$.
(P3) For every sequence $\{A_i\}_{i \in \mathbb{N}}$ of pairwise disjoint sets with $A_i \in \mathcal{P}(\Omega)$, $i \in \mathbb{N}$, the following applies:

© The Author(s), under exclusive license to Springer-Verlag GmbH, DE, part of
Springer Nature 2024
S. Schäffler, *Mathematics of Information*, Mathematics Study Resources 9,
https://doi.org/10.1007/978-3-662-69102-1_3

$$\mathbb{P}\left(\bigcup_{i=1}^{\infty} A_i\right) = \sum_{i=1}^{\infty} \mathbb{P}(A_i).$$

A subset $A \subseteq \Omega$ is referred to as an **event**; hence, the power set of Ω is also called the **set of events**. A single-element event $\{\omega\}$ is called an **elementary event**; thus, we always calculate the probability of events. In summary, we use the triple $(\Omega, \mathcal{P}(\Omega), \mathbb{P})$ as a mathematical object for a random experiment with at most countably many different results, which represents a special case of a probability space to be defined later.

Definition 3.1 (Discrete Probability Space) Let Ω be a non-empty at most countable set (so there is a subset $N \subseteq \mathbb{N}$ such that a bijection $N \to \Omega$ exists). Let further $\mathcal{P}(\Omega)$ be the power set (i.e., the set of all subsets) of Ω and let $\mathbb{P}$ be a mapping with the following properties:

(P1) $\mathbb{P} : \mathcal{P}(\Omega) \to [0, 1]$,
(P2) $\mathbb{P}(\emptyset) = 0, \mathbb{P}(\Omega) = 1$,
(P3) for every sequence $\{A_i\}_{i \in \mathbb{N}}$ of pairwise disjoint sets with $A_i \in \mathcal{P}(\Omega)$, $i \in \mathbb{N}$, the following holds:

$$\mathbb{P}\left(\bigcup_{i=1}^{\infty} A_i\right) = \sum_{i=1}^{\infty} \mathbb{P}(A_i),$$

then $(\Omega, \mathcal{P}(\Omega), \mathbb{P})$ is referred to as a **discrete probability space**.

Ω is referred to as the **set of outcomes**, $\mathcal{P}(\Omega)$ as the **set of events** and $\mathbb{P}$ as the **probability measure on $\mathcal{P}(\Omega)$**.
For $A \in \mathcal{P}(\Omega)$ the real number $\mathbb{P}(A)$ denotes the **probability for A**. ◁

Example 3.2 In roulette, one obtains for a game the result set

$$\Omega = \{0, 1, 2, \ldots, 35, 36\}.$$

Usually, the probabilities

$$\mathbb{P}(\{\omega\}) = \frac{1}{37} \quad \left(= \frac{1}{|\Omega|}\right), \quad \omega \in \Omega,$$

are set for the elementary events, where $|A|$ denotes the number of elements of a set A (cardinality of A). Thus, we obtain a probability measure on $\mathcal{P}(\Omega)$ by

$$\mathbb{P} : \mathcal{P}(\Omega) \to [0, 1], \quad A \mapsto \sum_{\omega \in A} \mathbb{P}(\{\omega\}) \quad \left(= \frac{|A|}{|\Omega|}\right).$$

Every player who takes note of the result of a game at the roulette table thereby receives the amount of information

$$I\left(\frac{1}{37}\right) = - \operatorname{ld}\left(\frac{1}{37}\right) = \operatorname{ld}(37) \approx 5.21 \text{ bit.}$$

Now, if we consider the result of eight games, one would choose

$$\Omega_8 = \{0, 1, 2, \ldots, 35, 36\}^8$$

and set the probabilities for the elementary events as follows:

$$\mathbb{P}(\{\omega\}) = \frac{1}{37^8} \quad \left(= \frac{1}{|\Omega_8|}\right), \quad \omega \in \Omega_8.$$

Thus, we obtain a probability measure on $\mathcal{P}(\Omega_8)$ by

$$\mathbb{P} : \mathcal{P}(\Omega_8) \to [0, 1], \quad A \mapsto \sum_{\omega \in A} \mathbb{P}(\{\omega\}) \quad \left(= \frac{|A|}{|\Omega_8|}\right).$$

Every player who takes note of the results of eight games at the roulette table thereby receives the amount of information

$$I\left(\frac{1}{37^8}\right) = - \operatorname{ld}\left(\frac{1}{37^8}\right) = 8 \cdot \operatorname{ld}(37) \approx 41.68 \text{ bit.}$$

Now, in roulette it is possible to bet on the event "even natural number", i.e., on the event $G := \{2, 4, 6, \ldots, 34, 36\}$. A player bets on the event G in each of the eight games and naturally wants to know with what probability he wins m times $(m = 0, 1, 2, \ldots, 7, 8)$. If one considers the mapping

$$X : \Omega_8 \to \{0, 1, 2, \ldots, 8\},$$

which counts how often an even natural number occurs in a tuple $\omega \in \Omega_8$, one obtains by

$$\mathbb{P}_X(\{i\}) := \mathbb{P}(\{\omega \in \Omega_8; \ X(\omega) = i\}) = \binom{8}{i}\left(\frac{18}{37}\right)^i\left(\frac{19}{37}\right)^{8-i}, \quad i = 0, \ldots, 8,$$

(binomial distribution) a probability measure $\mathbb{P}_X$ on $\mathcal{P}(\{0, 1, 2, \ldots, 7, 8\})$, where

$$\binom{8}{i} := \frac{8!}{(8 - i)! i!} \quad \text{(binomial coefficient).}$$

Now, if the player does not tell his wife the individual results of the eight games, but only how many of these eight games he has won, the following possible amounts of information arise for the wife:

$$I(\mathbb{P}_X(\{0\})) = -8 \cdot \mathrm{ld}\left(\frac{19}{37}\right) \approx 7.692 \text{ bit,}$$

$$I(\mathbb{P}_X(\{1\})) = -\mathrm{ld}(8) - \mathrm{ld}\left(\frac{18}{37}\right) - 7 \cdot \mathrm{ld}\left(\frac{19}{37}\right) \approx 4.770 \text{ bit,}$$

$$I(\mathbb{P}_X(\{2\})) = -\mathrm{ld}(28) - 2\,\mathrm{ld}\left(\frac{18}{37}\right) - 6 \cdot \mathrm{ld}\left(\frac{19}{37}\right) \approx 3.041 \text{ bit,}$$

$$I(\mathbb{P}_X(\{3\})) = -\mathrm{ld}(56) - 3\,\mathrm{ld}\left(\frac{18}{37}\right) - 5 \cdot \mathrm{ld}\left(\frac{19}{37}\right) \approx 2.119 \text{ bit,}$$

$$I(\mathbb{P}_X(\{4\})) = -\mathrm{ld}(70) - 4\,\mathrm{ld}\left(\frac{18}{37}\right) - 4 \cdot \mathrm{ld}\left(\frac{19}{37}\right) \approx 1.875 \text{ bit,}$$

$$I(\mathbb{P}_X(\{5\})) = -\mathrm{ld}(56) - 5\,\mathrm{ld}\left(\frac{18}{37}\right) - 3 \cdot \mathrm{ld}\left(\frac{19}{37}\right) \approx 2.275 \text{ bit,}$$

$$I(\mathbb{P}_X(\{6\})) = -\mathrm{ld}(28) - 6\,\mathrm{ld}\left(\frac{18}{37}\right) - 2 \cdot \mathrm{ld}\left(\frac{19}{37}\right) \approx 3.353 \text{ bit,}$$

$$I(\mathbb{P}_X(\{7\})) = -\mathrm{ld}(8) - 7\,\mathrm{ld}\left(\frac{18}{37}\right) - \mathrm{ld}\left(\frac{19}{37}\right) \approx 5.238 \text{ bit,}$$

$$I(\mathbb{P}_X(\{8\})) = -8 \cdot \mathrm{ld}\left(\frac{18}{37}\right) = 8.316 \text{ bit.}$$

◁

3.2   Average Amount of Information

In the second chapter, we calculated the average amount of information of a character according to Table 1.1 and compared it with the average word length of the coding given there. We now introduce the average amount of information for each discrete probability space, leading to a central size of information theory, which we will now define.

Definition 3.3 ((Shannon-)Entropy) Let $(\Omega, \mathcal{P}(\Omega), \mathbb{P})$ be a discrete probability space, then the average amount of information $\mathbb{S}_\mathbb{P}$ is given by

$$\mathbb{S}_\mathbb{P} := \sum_{\omega \in \Omega} \mathbb{P}(\{\omega\}) I(\mathbb{P}(\{\omega\})) \quad \left(= -\sum_{\omega \in \Omega} \mathbb{P}(\{\omega\}) \, \mathrm{ld}(\mathbb{P}(\{\omega\})) \right)$$

as **Entropy** or **Shannon Entropy** (named after CLAUDE ELWOOD SHANNON, the founder of information theory).   ◁

Example 3.4 Let's return to roulette (Example 3.1) and consider again the player who bets on the event

$$G = \{2, 4, 6, \ldots, 34, 36\}$$

in each of eight games. For eight games, a discrete probability space results with

$$\Omega_8 = \{0, 1, 2, \ldots, 35, 36\}^8$$

and

$$\mathbb{P}_8 : \mathcal{P}(\Omega_8) \to [0, 1], \quad A \mapsto \sum_{\omega \in A} \mathbb{P}(\{\omega\}) \quad \left(= \frac{|A|}{|\Omega_8|} \right).$$

Thus, we obtain the average amount of information

$$\mathbb{S}_{\mathbb{P}_8} = -\sum_{\omega \in \Omega_8} \mathbb{P}_8(\{\omega\}) \, \mathrm{ld}(\mathbb{P}_8(\{\omega\})) = -\mathrm{ld}\left(\frac{1}{37^8} \right) = 8 \cdot \mathrm{ld}(37) \approx 41.68 \text{ bit.}$$

Now, if the player does not tell his wife the individual results of the eight games, but only how many of these eight games he has won, a new discrete probability space is given to result with

$$\Omega_X = \{0, 1, 2, \ldots, 7, 8\}$$

and

$$\mathbb{P}_X : \mathcal{P}(\Omega_X) \to [0, 1], \quad A \mapsto \sum_{i \in A} \binom{8}{i} \left(\frac{18}{37} \right)^i \left(\frac{19}{37} \right)^{8-i}.$$

For the entropy, we obtain

$$\mathbb{S}_{\mathbb{P}_X} = -\sum_{i=0}^{8} \mathbb{P}_X(\{i\}) \, \mathrm{ld}(\mathbb{P}_X(\{i\})) \approx 2.5437 \text{ bit.}$$

Since from a result of the eight roulette games, one can always determine the number of games in which the event G occurred, but conversely, from knowing how often the event G occurred in eight games, one cannot infer the results of the eight games, one intuitively expects that the entropy $\mathbb{S}_{\mathbb{P}_8}$ is greater than $\mathbb{S}_{\mathbb{P}_X}$, which is confirmed by calculation. Now let's assume that the player informs his wife via SMS, then the following binary coding

$$0 \to 0 \qquad 1 \to 1 \qquad 2 \to 10$$
$$3 \to 11 \qquad 4 \to 100 \qquad 5 \to 101$$
$$6 \to 110 \qquad 7 \to 111 \qquad 8 \to 1000$$

with an average word length of approximately 2.5684 bits would initially be offered. However, if the player plays several series of eight and tells his wife, for example, by "200" that he won twice in the first series and did not win anything in the second and third series, then the binary representation of the three digits 2,0,0 of "200" would be 1000, which could also mean that a series with eight wins or two series with four wins in the first and no wins in the second series were played. However, if one uses, for example, the binary coding

$$0 \to 1110001 \quad 1 \to 11101 \quad 2 \to 110$$
$$3 \to 01 \qquad\quad 4 \to 10 \qquad 5 \to 00$$
$$6 \to 1111 \qquad 7 \to 111001 \quad 8 \to 1110000$$

with an average word length of approximately 2.5832 bits, misunderstandings of this kind are excluded.   ◁

In the next section, we will get to know an algorithm that provides the uniquely reconstructable coding for the numbers $0, 1, \ldots, 7, 8$ and a coding like in Table 1.1 as indicated in the above example.

If one has a discrete probability space $(\Omega, \mathcal{P}(\Omega), \mathbb{P})$, a non-empty countable set Ω_X and a mapping

$$X : \Omega \to \Omega_X,$$

one obtains via

$$\mathbb{P}_X : \mathcal{P}(\Omega_X) \to [0, 1], \quad A \mapsto \mathbb{P}(\{\omega \in \Omega; \ X(\omega) \in A\})$$

a discrete probability space $(\Omega_X, \mathcal{P}(\Omega_X), \mathbb{P}_X)$. As the following theorem shows, the entropy does not increase when transitioning from $(\Omega, \mathcal{P}(\Omega), \mathbb{P})$ to $(\Omega_X, \mathcal{P}(\Omega_X), \mathbb{P}_X)$.

Theorem and Definition 3.5 (Entropy of the Image Measure) *Given a discrete probability space* $(\Omega, \mathcal{P}(\Omega), \mathbb{P})$, *a non-empty countable set* Ω_X *and a mapping*

$$X : \Omega \to \Omega_X,$$

$$\mathbb{P}_X : \mathcal{P}(\Omega_X) \to [0, 1], \quad A \mapsto \mathbb{P}(\{\omega \in \Omega; \ X(\omega) \in A\})$$

is a probability measure and it holds:

$$\mathbb{S}_\mathbb{P} \geq \mathbb{S}_{\mathbb{P}_X}.$$

The probability measure $\mathbb{P}_X$ *is referred to as the* **image measure** *of* $\mathbb{P}$ *under* X.   ◁

Let $\{B_i\}_{i\in\mathbb{N}}$ be a sequence of pairwise disjoint sets with $B_i \in \mathcal{P}(\Omega_X), i \in \mathbb{N}$:

$$\mathbb{P}_X(\emptyset) = \mathbb{P}(\{\omega \in \Omega; \ X(\omega) \in \emptyset\}) = \mathbb{P}(\emptyset) = 0$$
$$\mathbb{P}_X(\Omega_X) = \mathbb{P}(\{\omega \in \Omega; \ X(\omega) \in \Omega_X\}) = \mathbb{P}(\Omega) = 1$$

$$\mathbb{P}_X\left(\bigcup_{i=1}^{\infty} B_i\right) = \mathbb{P}\left(\left\{\omega \in \Omega; \ X(\omega) \in \bigcup_{i=1}^{\infty} B_i\right\}\right)$$

$$= \mathbb{P}\left(\bigcup_{i=1}^{\infty} \{\omega \in \Omega; X(\omega) \in B_i\}\right)$$

$$= \sum_{i=1}^{\infty} \mathbb{P}(\{\omega \in \Omega; X(\omega) \in B_i\}) = \sum_{i=1}^{\infty} \mathbb{P}_X(B_i),$$

since the sets $\{\omega \in \Omega; X(\omega) \in B_i\}$, $i \in \mathbb{N}$, are pairwise disjoint.

$$\mathbb{S}_{\mathbb{P}} = -\sum_{\omega \in \Omega} \mathbb{P}(\{\omega\})\, \mathrm{ld}\, (\mathbb{P}(\{\omega\})) = -\sum_{\omega_X \in \Omega_X}\ \sum_{\{\omega \in \Omega; X(\omega) = \omega_X\}} \mathbb{P}(\{\omega\})\, \mathrm{ld}\, (\mathbb{P}(\{\omega\}))$$

$$\geq -\sum_{\omega_X \in \Omega_X}\ \sum_{\{\omega \in \Omega; X(\omega) = \omega_X\}} \mathbb{P}(\{\omega\})\, \mathrm{ld}\, (\mathbb{P}_X(\{\omega_X\}))$$

$$= -\sum_{\omega_X \in \Omega_X} \mathrm{ld}\, (\mathbb{P}_X(\{\omega_X\})) \sum_{\{\omega \in \Omega; X(\omega) = \omega_X\}} \mathbb{P}(\{\omega\})$$

$$= -\sum_{\omega_X \in \Omega_X} \mathrm{ld}\, (\mathbb{P}_X(\{\omega_X\})) \mathbb{P}_X(\{\omega_X\})$$

$$= \mathbb{S}_{\mathbb{P}_X}.$$

q.e.d

In the following theorem, we investigate the maximum entropy for finite outcome sets.

▶ **Theorem 3.6 (Maximum Entropy for Finite Outcome Sets)** *Let Ω be a non-empty set with $k \in \mathbb{N}$ elements, then for every probability measure $\mathbb{P}$ on $\mathcal{P}(\Omega)$ holds:*

$$\mathbb{S}_{\mathbb{P}} \leq \mathrm{ld}(k).$$

Equality holds exactly when

$$\mathbb{P}(\{\omega\}) = \frac{1}{k} \quad \text{for all} \quad \omega \in \Omega.$$

If one considers the function

$$f : [0, 1] \to \mathbb{R}, \quad x \mapsto \begin{cases} 0 & \text{if } x = 0 \\ x\, \mathrm{ld}(x) & \text{if } x \neq 0 \end{cases},$$

then f is strictly convex. With Jensen's inequality it follows:

$$f\left(\frac{1}{k}\sum_{i=1}^{k} x_k\right) \leq \frac{1}{k}\sum_{i=1}^{k} f(x_k), \quad x_1, \ldots, x_k \in [0, 1],$$

where equality holds exactly when $x_1 = x_2 = \ldots = x_k$.

   With

$$\Omega = \{\omega_1, \ldots, \omega_k\}$$

we now set

$$x_i = \mathbb{P}(\{\omega_i\}), \quad i = 1, \ldots, k$$

and obtain

$$\frac{1}{k} \operatorname{ld}\left(\frac{1}{k}\right) \leq \frac{1}{k} \sum_{i=1}^{k} \mathbb{P}(\{\omega_i\}) \operatorname{ld}(\mathbb{P}(\{\omega_i\}))$$

or

$$\operatorname{ld}(k) \geq -\sum_{i=1}^{k} \mathbb{P}(\{\omega_i\}) \operatorname{ld}(\mathbb{P}(\{\omega_i\})),$$

where equality holds exactly when

$$\mathbb{P}(\{\omega_i\}) = \frac{1}{k}, \quad i = 1, \ldots, k.$$

q.e.d

In the following example, we consider a random experiment with $\Omega = \mathbb{N}$ and finite entropy.

Example 3.7 For a coin, it is assumed that "heads" occurs with probability $p \in (0, 1)$ and "tails" occurs with probability $(1 - p)$. If we consider a random experiment where the coin is tossed until "heads" appears for the first time and the number of tosses required for this is recorded as the result, then $\Omega = \mathbb{N}$ and by

$$\mathbb{P}(\{i\}) = p(1 - p)^{i-1}, \quad i \in \mathbb{N} \quad \text{(geometric distribution)}$$

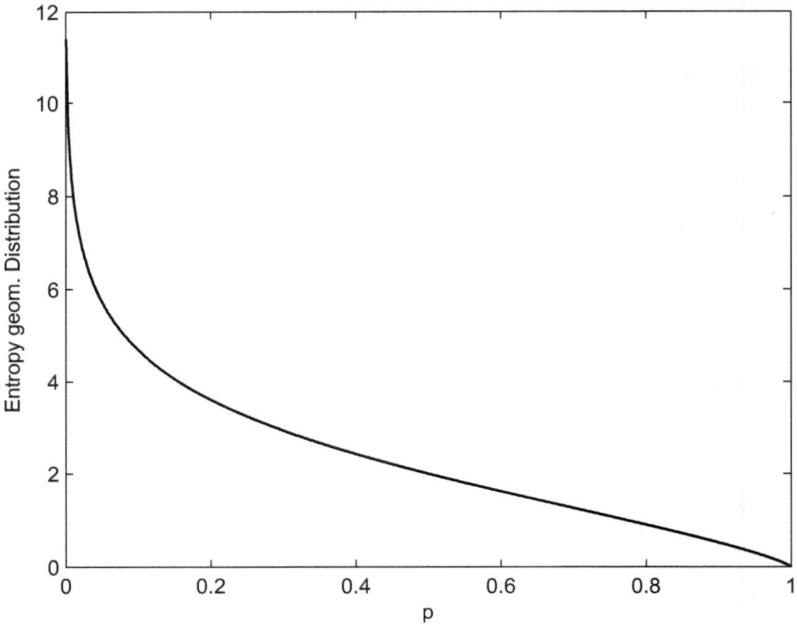

Fig. 3.1 The entropy of the geometric distribution

the corresponding probability measure on $\mathcal{P}(\mathbb{N})$ is given. For the entropy, we obtain:

$$\mathbb{S}_\mathbb{P} = -\sum_{i=1}^{\infty} p(1-p)^{i-1}\,\mathrm{ld}\left(p(1-p)^{i-1}\right) =$$

$$= -p\,\mathrm{ld}(p)\sum_{i=1}^{\infty}(1-p)^{i-1} - p\,\mathrm{ld}(1-p)\sum_{i=1}^{\infty}(i-1)(1-p)^{i-1}$$

$$= -\mathrm{ld}(p) - p\,\mathrm{ld}(1-p)\frac{1-p}{p^2} = -\mathrm{ld}(p) - \mathrm{ld}(1-p)\frac{(1-p)}{p}.$$

If the coin is fair $\left(p = \frac{1}{2}\right)$, the entropy is equal to 2 bit. Furthermore, (Fig. 3.1):

$$\lim_{p\to 0}\mathbb{S}_\mathbb{P} = \infty, \quad \lim_{p\to 1}\mathbb{S}_\mathbb{P} = 0.$$

◁

3.3 Huffman Coding

In 1952, DAVID A. HUFFMAN (1925–1999) proposed a coding of probabilities, in which no codeword can be the beginning of another codeword (unique reconstructability) and which is optimal under this condition (i.e., with minimal average word length). We demonstrate Huffman coding using the example of the roulette player who bets on the event $G = \{2, 4, 6, \ldots, 34, 36\}$ in eight games and informs his wife of the number of games won via a text message. From Example 3.1 we know the following probabilities, sorted by size, where the index denotes the corresponding number of games won (if two or more probabilities are the same, the order of the same probabilities is chosen randomly):

$$0.27264_4$$
$$0.23023_3$$
$$0.20663_5$$
$$0.12151_2$$
$$0.09788_6$$
$$0.03665_1$$
$$0.02649_7$$
$$0.00484_0$$
$$0.00313_8$$

In the first step, we add the two smallest probabilities and note in the tuple $\tau_1 = (0, 8)$ the two results (more precisely: elementary events) that belong to the added probabilities; the results are noted in the order of decreasing associated probabilities (so here 0 before 8; if probabilities are the same, one chooses randomly). The result is $0.00797_{\tau_1=(0,8)}$. This new probability is now inserted into the above list at the appropriate place (sorted by size) and the two added probabilities are removed. The following probabilities result:

$$0.27264_4$$
$$0.23023_3$$
$$0.20663_5$$
$$0.12151_2$$
$$0.09788_6$$
$$0.03665_1$$
$$0.02649_7$$
$$0.00797_{\tau_1=(0,8)}.$$

If one again adds the two smallest probabilities and note in the tuple $\tau_2 = (7, \tau_1)$ the two indices, which—in order of size—belong to the added probabilities, one gets:

$$0.27264_4$$
$$0.23023_3$$
$$0.20663_5$$
$$0.12151_2$$
$$0.09788_6$$
$$0.03446_{\tau_2=(7,\tau_1)}.$$

This procedure is now continued until only the probability 1 remains:

0.27264_4	0.27264_4	$0.29050_{\tau_5=(\tau_4,2)}$	$0.43686_{\tau_6=(3,5)}$
0.23023_3	0.23023_3	0.27264_4	0.29050_{τ_5}
0.20663_5	0.20663_5	0.23023_3	0.27264_4
0.12151_2	$0.16899_{\tau_4=(6,\tau_3)}$	0.20663_5	
0.09788_6	0.12151_2		
$0.07111_{\tau_3=(1,\tau_2)}$			

and

$$0.56314_{\tau_7=(\tau_5,4)} \qquad 1_{\tau_8=(\tau_7,\tau_6)}$$
$$0.43686_{\tau_6}$$

Now the assignment of binary characters to the results begins

$$0, 1, 2, \ldots, 7, 8.$$

One starts with the tuple that has the largest index, so $\tau_8 = (\tau_7, \tau_6)$. The first element of the tuple is assigned the bit 1, the second the bit 0, so

$$1 \to \tau_7 \quad 0 \to \tau_6.$$

If you continue with $\tau_7 = (\tau_5, 4)$, the first element τ_5 is again assigned 1, the second (i.e., the result 4) the bit 0. But since τ_7 has already been assigned the bit 1, the assignments for τ_5 and for the result 4 are appended to the bit 1:

$$11 \to \tau_5 \quad 10 \to 4.$$

Since $\tau_6 = (3, 5)$, the first element (i.e., the result 3) is again assigned the bit 1, the second (i.e., the result 5) the bit 0. But since τ_6 has already been assigned the bit 0, the assignments for the result 3 and for the result 5 are appended to the bit 0:

$$01 \to 3 \quad 00 \to 5.$$

With $\tau_5 = (\tau_4, 2)$, τ_4 is assigned the bit 1, the second element (i.e., the result 2) the bit 0. But sinceτ_5 Since the bit sequence 11 was already assigned, the assignments for τ_4 and for the result 2 of the bit sequence 11 are appended:

$$111 \to \tau_4 \quad 110 \to 2.$$

With $\tau_4 = (6, \tau_3)$ we thus obtain

$$1111 \to 6 \quad 1110 \to \tau_3$$

and due to $\tau_3 = (1, \tau_2)$:

$$11101 \to 1 \quad 11100 \to \tau_2.$$

From $\tau_2 = (7, \tau_1)$ follows

$$111001 \to 7 \quad 111000 \to \tau_1$$

and finally due to $\tau_1 = (0, 8)$:

$$1110001 \to 0 \quad 1110000 \to 8.$$

In summary, we thus obtain the assignment given in example 3.2:

$$0 \to 1110001 \quad 1 \to 11101 \quad 2 \to 110$$
$$3 \to 01 \quad\quad 4 \to 10 \quad\quad 5 \to 00$$
$$6 \to 1111 \quad\quad 7 \to 111001 \quad 8 \to 1110000.$$

One could have also assigned the bit 0 to the first element of a tuple and the bit 1 to the second element; however, once a strategy is chosen, it must be consistently applied to all tuples.

The idea of David Huffman can not only be used for binary coding, but works much more generally. Let's stick with the above example with nine probabilities and now assume that the character set for coding does not consist of $\{0, 1\}$, but of $\{a, b, c, d\}$. Since we now have four characters, the four smallest probabilities must always be added. So, from nine probabilities, we get six probabilities after the first step and finally three probabilities. In order to always be able to add four probabilities, one probability is missing at the end. We correct this by introducing an artificial new result ● with probability zero:

$$0.27264_4$$
$$0.23023_3$$
$$0.20663_5$$
$$0.12151_2$$
$$0.09788_6$$
$$0.03665_1$$
$$0.02649_7$$
$$0.00484_0$$
$$0.00313_8$$
$$0_●.$$

Hence, with a character set with k characters for coding, a maximum of $k - 2$ new results with probability zero must be added. The first step now results in

$$0.27264_4$$
$$0.23023_3$$
$$0.20663_5$$
$$0.12151_2$$
$$0.09788_6$$
$$0.03665_1$$
$$0.03446_{\tau_1=(7,0,8,●)},$$

where we have ordered the results in $\tau_1 = (7, 0, 8, ●)$ by decreasing probabilities. Step two yields:

$$0.29050_{\tau_2=(2,6,1,\tau_1)}$$
$$0.27264_4$$
$$0.23023_3$$
$$0.20663_5$$

and finally

$$1_{\tau_3=(\tau_2,4,3,5)}.$$

If we now establish the strategy that, for example, the symbol b is assigned to the first element of a tuple τ, the symbol c to the second, the symbol a to the third element and the symbol d to the fourth (which now has to happen for all tuples), the following results:

$$b \to \tau_2 \quad c \to 4 \quad a \to 3 \quad d \to 5$$

and because of $\tau_2 = (2, 6, 1, \tau_1)$:

$$bb \rightarrow 2 \quad bc \rightarrow 6 \quad ba \rightarrow 1 \quad bd \rightarrow \tau_1.$$

Because of $\tau_1 = (7, 0, 8, \bullet)$ we finally get

$$bdb \rightarrow 7 \quad bdc \rightarrow 0 \quad bda \rightarrow 8,$$

so in total:

$$0 \rightarrow bdc \quad 1 \rightarrow ba \quad 2 \rightarrow bb$$
$$3 \rightarrow a \quad\quad 4 \rightarrow c \quad\; 5 \rightarrow d$$
$$6 \rightarrow bc \quad\; 7 \rightarrow bdb \quad 8 \rightarrow bda.$$

The average word length is now about 1.32496. It makes no sense to compare this value with the average amount of information of about 2.5437 bits, since four characters are available for coding, while the function I for measuring the amount of information was chosen to be the logarithm to the base 2. If the function I had been chosen to be the base 4, which corresponds to the number of available characters for coding, our example would yield an amount of information of about 1.27185, which provides a meaningful comparison value to the average word length. A precise formulation and analysis of Huffman coding can be found, for example, in [HeiQua95].

The Maximum Entropy Principle

4

4.1 Maximum Average Information under Constraints

In Theorem 3.6 it was shown that for a non-empty finite result set Ω, the probability measure on $\mathcal{P}(\Omega)$ with maximum entropy is given by

$$\mathbb{P}(\{\omega\}) = \frac{1}{|\Omega|}, \quad \omega \in \Omega \quad (|\Omega| \text{ number of elements of } \Omega).$$

Now we investigate the same question under constraints.

Let $n \in \mathbb{N}$ and

$$f_{\mathbb{S}} : [0,1]^n \to \mathbb{R}_0^+, \quad \mathbf{x} = (x_1, \ldots, x_n) \mapsto -\sum_{j \in J_{\mathbf{x}}} x_j \, \mathrm{ld}(x_j)$$

with

$$J_{\mathbf{x}} = \{k \in \{1, \ldots, n\}; \ x_k > 0\},$$

then $f_{\mathbb{S}}$ due to

$$\lim_{x \to 0} x \, \mathrm{ld}(x) = 0$$

is continuous and strictly concave on $[0,1]^n$ (see Theorem 2.1 and Fig. 4.1).

We consider the maximization problem

$$\max_{\mathbf{p}=(p_1,\ldots,p_n)} \left\{ f_{\mathbb{S}}(\mathbf{p}); \quad p_i \geq 0, \quad i = 1, \ldots, n, \right.$$

$$\sum_{i=1}^{n} p_i = 1,$$

$$\left. \mathbf{c}_r^\top \mathbf{p} = b_r, \quad r = 1, \ldots, m \right\}$$

© The Author(s), under exclusive license to Springer-Verlag GmbH, DE, part of
Springer Nature 2024
S. Schäffler, *Mathematics of Information*, Mathematics Study Resources 9,
https://doi.org/10.1007/978-3-662-69102-1_4

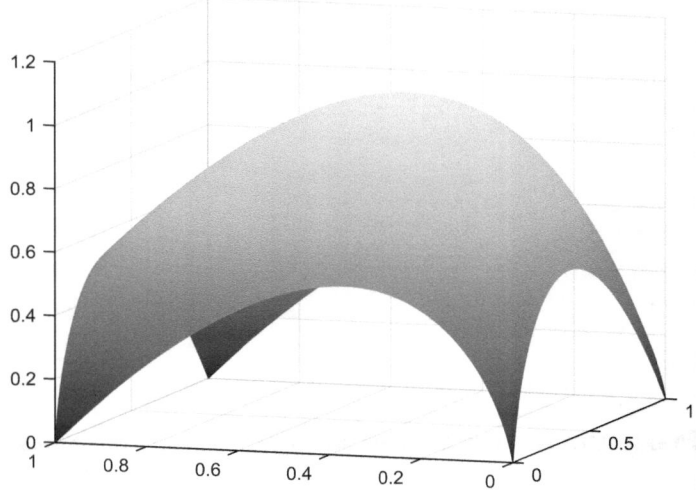

Fig. 4.1 f_S with definition range $[0, 1]^2$

with $\mathbf{c}_r \in \mathbb{R}^n, r = 1, \dots, m$. The feasible region

$$R := \left\{ \quad p_i \geq 0, \quad i = 1, \dots, n, \right.$$

$$\sum_{i=1}^{n} p_i = 1,$$

$$\left. \mathbf{c}_r^\top \mathbf{p} = b_r, \quad r = 1, \dots, m \right\}$$

of this maximization problem is either empty, consists of one point, or consists of infinitely many points and forms a convex and compact set. If R is not empty, our maximization problem with a strictly concave objective function always has a unique solution $\hat{\mathbf{p}}$. Let now $I \subseteq \{1, 2, \dots, n\}$ be the set of all indices i with $\hat{p}_i > 0$, the components $\hat{p}_i, i \in I$, of $\hat{\mathbf{p}}$ are given by the unique solution of the nonlinear system of equations (Lagrange approach):

$$- \mathrm{ld}(p_i) - \frac{1}{\ln(2)} + \sum_{r=1}^{m} \lambda_r(\mathbf{c}_r)_i + \lambda_{m+1} = 0, \quad i \in I$$

$$\sum_{i \in I} (\mathbf{c}_1)_i p_i - b_1 = 0$$

$$\vdots$$

$$\sum_{i \in I} (\mathbf{c}_m)_i p_i - b_m = 0$$

$$\sum_{i \in I} p_i - 1 = 0$$

in the variables p_i, $i \in I$, $\lambda_1, \ldots, \lambda_{m+1}$. We obtain:

$$p_i = \exp \left(\lambda_{m+1} \ln(2) - 1 + \ln(2) \sum_{r=1}^{m} \lambda_r(\mathbf{c}_r)_i \right), \quad i \in I$$

$$\sum_{i \in I} (\mathbf{c}_1)_i p_i - b_1 = 0$$

$$\vdots$$

$$\sum_{i \in I} (\mathbf{c}_m)_i p_i - b_m = 0$$

$$\sum_{i \in I} p_i - 1 = 0$$

or

$$\lambda_{m+1} = \frac{1 - \ln \left(\sum_{i \in I} \exp \left(\ln(2) \sum_{r=1}^{m} \lambda_r(\mathbf{c}_r)_i \right) \right)}{\ln(2)}$$

$$p_i = \frac{2^{\sum_{r=1}^{m} \lambda_r(\mathbf{c}_r)_i}}{\sum_{i \in I} 2^{\sum_{r=1}^{m} \lambda_r(\mathbf{c}_r)_i}}, \quad i \in I$$

$$\sum_{i \in I} (\mathbf{c}_1)_i p_i - b_1 = 0$$

$$\vdots$$

$$\sum_{i \in I} (\mathbf{c}_m)_i p_i - b_m = 0.$$

Finally, the unique solution $\hat{\lambda}$ of the nonlinear system of equations in $\lambda_1, \ldots, \lambda_m$:

$$\sum_{i \in I} (\mathbf{c}_1)_i \frac{2^{\sum_{r=1}^{m} \lambda_r (\mathbf{c}_r)_i}}{\sum_{i \in I} 2^{\sum_{r=1}^{m} \lambda_r (\mathbf{c}_r)_i}} - b_1 = 0$$

$$\vdots$$

$$\sum_{i \in I} (\mathbf{c}_m)_i \frac{2^{\sum_{r=1}^{m} \lambda_r (\mathbf{c}_r)_i}}{\sum_{i \in I} 2^{\sum_{r=1}^{m} \lambda_r (\mathbf{c}_r)_i}} - b_m = 0$$

remains to be determined.

Example 4.1 Let $\Omega = \{1, 2, 3, 4, 5, 6\}$ be the result set when rolling a die once. For which probability measure $\mathbb{P}$ on $\mathcal{P}(\Omega)$ is the entropy maximal when the averaged result μ is determined by

$$\mu := \sum_{i=1}^{6} i \cdot \mathbb{P}(\{i\}) = 3.$$

Hence, the maximization problem

$$\max_{\mathbf{p}=(p_1,\ldots,p_6)} \left\{ f_{\mathbb{S}}(\mathbf{p}); \quad p_i \geq 0, \quad i = 1, \ldots, 6, \right.$$

$$\sum_{i=1}^{6} p_i = 1,$$

$$\left. \sum_{i=1}^{6} i \cdot p_i = 3 \right\}$$

needs to be solved. Let's first assume that all components of the unique solution $\hat{\mathbf{p}}$ are greater than zero, then we get:

$$\hat{p}_i = \frac{2^{\hat{\lambda} \cdot i}}{\sum_{i=1}^{6} 2^{\hat{\lambda} \cdot i}}, \quad i = 1, \ldots, 6,$$

where for the calculation of $\hat{\lambda}$ the nonlinear equation

$$\sum_{i=1}^{6} i \cdot \frac{2^{\lambda \cdot i}}{\sum\limits_{i=1}^{6} 2^{\lambda \cdot i}} - 3 = 0$$

in λ needs to be solved.

If one sets $x := 2^{\lambda}$, a solution $\hat{x} > 0$ of the nonlinear equation

$$3x^5 + 2x^4 + x^3 - x - 2 = 0$$

is sought. Since the function

$$h : [0, \infty) \to \mathbb{R}, \quad x \mapsto 3x^5 + 2x^4 + x^3 - x - 2$$

is strictly convex, it follows from $h(0) = -2$ and $h(1) = 3$ the existence of a unique solution $\hat{x} > 0$ of

$$3x^5 + 2x^4 + x^3 - x - 2 = 0$$

and thus the existence of a unique solution $\hat{\lambda}$ of

$$\sum_{i=1}^{6} i \cdot \frac{2^{\lambda \cdot i}}{\sum\limits_{i=1}^{6} 2^{\lambda \cdot i}} - 3 = 0.$$

By a simple bisection method, for the nonlinear equation

$$3x^5 + 2x^4 + x^3 - x - 2 = 0$$

the approximation

$$\hat{x} \approx 0.84 \quad (\text{also } \hat{\lambda} \approx -0.2515)$$

of the sought solution is obtained.

To therefore achieve the average result of $\mu = 3$ when rolling a dice, one would have to prepare the dice for maximum entropy such that

$$\mathbb{P}(\{1\}) \approx 0.247 \quad \mathbb{P}(\{2\}) \approx 0.207 \quad \mathbb{P}(\{3\}) \approx 0.174$$
$$\mathbb{P}(\{4\}) \approx 0.146 \quad \mathbb{P}(\{5\}) \approx 0.123 \quad \mathbb{P}(\{6\}) \approx 0.103$$

applies. For $\mu = 3.5$ we would have $\mathbb{P}(\{i\}) = \frac{1}{6}$ for $i = 1, \ldots, 6$.   ◁

An important application of the maximum entropy method arises in statistical physics.

4.2   Statistical Physics

In a hollow body with a given volume, there is a fixed number N of molecules. We consider this thermodynamic system to be closed; thus, there is no interaction with the environment of the cavity. Each molecule in this space possesses an energy (the so-called **internal energy**), which is given by its mechanical properties (mass, speed). An exchange of internal energy occurs through the collision of two molecules. However, the sum E_{sum} of the internal energy of all molecules (and thus the average energy $E_M = E_{sum}/N$ per molecule) remains constant as a result of the system's closure. Each molecule can also only assume an integer multiple $k \cdot E$, $k \in \mathbb{N}$, of an energy amount E as an internal energy level. According to the main principles of thermodynamics, an exchange of internal energy between the molecules continues until a thermodynamic equilibrium state is reached (see [Stier10]). Let p_k be the probability that a molecule assumes the internal energy levelkE, $k \in \mathbb{N}$, then the entropy

$$-\sum_{k=1}^{\infty} p_k \, \mathrm{ld}(p_k)$$

(where again $0 \cdot \mathrm{ld}(0) = 0$ should apply) can be examined under the condition

$$\sum_{k=1}^{\infty} p_k kE = E_M \quad \text{(constant average energy per molecule)};$$

this obviously only makes sense for

$$0 < E \le E_M.$$

If you choose $E = E_M$, you immediately get

$$p_1 = 1, \quad p_k = 0, \ k \in \mathbb{N} \setminus \{1\}.$$

Therefore, we now start from

$$0 < E < E_M.$$

The thermodynamic equilibrium state is characterized by the solution of the following maximization problem

$$\max_{\{p_k\}_{k \in \mathbb{N}}} \left\{ -\sum_{k=1}^{\infty} p_k \, \mathrm{ld}(p_k); \qquad p_k \ge 0, \ k \in \mathbb{N}, \right.$$

$$\sum_{k=1}^{\infty} p_k = 1,$$

$$\left. \sum_{k=1}^{\infty} p_k kE = E_M \right\}$$

Analogous to the case with a finite number of variables, we obtain the unique solution

$$\hat{p}_k = \frac{2^{\hat{\lambda}kE}}{\sum\limits_{k=1}^{\infty} 2^{\hat{\lambda}kE}}, \quad k \in \mathbb{N},$$

if $\hat{\lambda} < 0$ represents the solution of the nonlinear equation

$$\sum_{k=1}^{\infty} \frac{2^{\hat{\lambda}kE}}{\sum\limits_{k=1}^{\infty} 2^{\hat{\lambda}kE}} kE = E_M,$$

because only $\hat{\lambda} < 0$ guarantees

$$\sum_{k=1}^{\infty} 2^{\hat{\lambda}kE} < \infty.$$

Now let $x := 2^{\hat{\lambda}E}$, then a solution $\hat{x} > 0$ of the nonlinear equation

$$\frac{E \sum\limits_{k=1}^{\infty} x^k k}{\sum\limits_{k=1}^{\infty} x^k} = E_M$$

is sought; for this solution,

$$\hat{x} < 1 \quad \text{because} \quad \hat{\lambda} < 0$$

must also apply. Since

$$\sum_{k=1}^{\infty} x^k = \frac{x}{1-x} \quad \text{and} \quad \sum_{k=1}^{\infty} x^k k = \frac{x}{(1-x)^2} \quad \text{with} \quad 0 < x < 1,$$

the nonlinear equation

$$E \frac{x}{(1-x)^2} = E_M \frac{x}{1-x} \quad \text{with} \quad 0 < x < 1$$

is to be solved. It results in

$$\hat{x} = \frac{E_M - E}{E_M} \quad \text{and therefore} \quad \hat{\lambda} = \frac{1}{E} \operatorname{ld}\left(\frac{E_M - E}{E_M}\right) < 0.$$

For the maximum entropy we obtain:

$$
\mathbb{S}_{\max} = -\sum_{k=1}^{\infty} \hat{p}_k \, \mathrm{ld}(\hat{p}_k) = -\sum_{k=1}^{\infty} \frac{2^{\hat{\lambda}kE}}{\sum\limits_{k=1}^{\infty} 2^{\hat{\lambda}kE}} \, \mathrm{ld}\left(\frac{2^{\hat{\lambda}kE}}{\sum\limits_{k=1}^{\infty} 2^{\hat{\lambda}kE}} \right)
$$

$$
= -\sum_{k=1}^{\infty} \frac{2^{\hat{\lambda}kE}}{\sum\limits_{k=1}^{\infty} 2^{\hat{\lambda}kE}} \, \mathrm{ld}\left(2^{\hat{\lambda}kE} \right) + \sum_{k=1}^{\infty} \frac{2^{\hat{\lambda}kE}}{\sum\limits_{k=1}^{\infty} 2^{\hat{\lambda}kE}} \, \mathrm{ld}\left(\sum_{k=1}^{\infty} 2^{\hat{\lambda}kE} \right)
$$

$$
= -\hat{\lambda} \sum_{k=1}^{\infty} \frac{2^{\hat{\lambda}kE}}{\sum\limits_{k=1}^{\infty} 2^{\hat{\lambda}kE}} kE + \mathrm{ld}\left(\sum_{k=1}^{\infty} 2^{\hat{\lambda}kE} \right) = -\hat{\lambda} E_M + \mathrm{ld}\left(\frac{2^{\hat{\lambda}E}}{1 - 2^{\hat{\lambda}E}} \right)
$$

$$
= -\hat{\lambda} E_M + \hat{\lambda} E - \mathrm{ld}\left(1 - 2^{\hat{\lambda}E} \right)
$$

$$
= -\frac{E_M}{E} \, \mathrm{ld}\left(\frac{E_M - E}{E_M} \right) + \mathrm{ld}\left(\frac{E_M - E}{E_M} \right) - \mathrm{ld}\left(1 - 2^{\mathrm{ld}\left(\frac{E_M - E}{E_M} \right)} \right)
$$

$$
= -\frac{E_M}{E} \, \mathrm{ld}\left(\frac{E_M - E}{E_M} \right) + \mathrm{ld}\left(\frac{E_M - E}{E_M} \right) - \mathrm{ld}\left(\frac{E}{E_M} \right)
$$

$$
= -\mathrm{ld}\left(1 - \frac{E}{E_M} \right) \frac{1 - \frac{E}{E_M}}{\frac{E}{E_M}} - \mathrm{ld}\left(\frac{E}{E_M} \right)
$$

$$
= -\mathrm{ld}\left(1 - e \right) \frac{1 - e}{e} - \mathrm{ld}\left(e \right) \quad \text{with } e = \frac{E}{E_M}.
$$

But this is precisely the entropy of the geometric distribution from example 3.2 with $p = e$.

This result is now to be interpreted physically. Assuming that energy is measured in Joule [J] as usual, the temperature T in Kelvin [K] in the equilibrium state is given by $-\hat{\lambda}$ through

$$
-\hat{\lambda} = \frac{1}{\beta T},
$$

where

$$
\beta \approx 1.3806505 \cdot 10^{-23} \frac{\text{J}}{\text{K}} \quad \text{(Boltzmann constant)}.
$$

Thus, the quantity $-\hat{\lambda}$ is given in the physical unit $\left[\frac{1}{\text{J}} \right]$ and the calculated probabilities represent—as expected—no physical quantities. The equation

$$
\mathbb{S}_{\max} = -\hat{\lambda} E_M + \hat{\lambda} E - \mathrm{ld}\left(1 - 2^{\hat{\lambda}E} \right)
$$

can now be written in the form

$$\underbrace{NE + N\beta T \operatorname{ld}\left(1 - 2^{-\frac{E}{\beta T}}\right)}_{=:F} = E_{\text{sum}} - N\beta T \mathbb{S}_{\text{max}}$$

common for thermodynamics. The quantity F is referred to as **free energy**. The total energy E_{sum} is composed of the energy $N\beta T\mathbb{S}_{\text{max}}$ (heat energy) and the free energy F, which indicates how much mechanical work the system performs in the equilibrium state (for example, on the inner walls of the cavity); a measure of this is the product of pressure and volume.

The fact that nature brings about an equilibrium state with maximum entropy can be interpreted information-theoretically as follows: The average amount of information one receives when randomly selecting a molecule in the equilibrium state and considering its internal energy as a result is maximal among all possible states of the closed system.

The second law of thermodynamics now states that during the time that passes until nature has maximized the entropy, it cannot become smaller. The maximization thus takes place through monotonically increasing entropies over time.

Conditional Probabilities

5

5.1 Sufficiency

Starting from a discrete probability space $(\Omega, \mathcal{P}(\Omega), \mathbb{P})$ and a set $B \subseteq \Omega$ with $\mathbb{P}(B) > 0$ one obtains via

$$\mathbb{P}^B : \mathcal{P}(\Omega) \to [0, 1], \quad A \mapsto \frac{\mathbb{P}(A \cap B)}{\mathbb{P}(B)}$$

another probability measure on $\mathcal{P}(\Omega)$. Since $\mathbb{P}^B(B) = 1$, one interprets the probability $\mathbb{P}^B(A)$ as the probability of the event A given that the event B definitely occurs (conditional probability). One has, so to speak, reduced the set of possible outcomes Ω to the set B. If now $\mathbb{P}^B(A) = \mathbb{P}(A)$ applies, then $\mathbb{P}(A \cap B) = \mathbb{P}(A) \cdot \mathbb{P}(B)$ follows; in this case, the probability for A is not affected by the reduction of the set of outcomes from Ω to B; one says, the events A and B are **stochastically independent**. Therefore, when introducing the function I to measure the amount of information, we also required

$$I(pq) = I(p) + I(q);$$

if the "message" A occurs with probability p, the message B with probability q and both messages with probability pq, then the respective amounts of information should add up if both messages occur (since stochastic independence is present).

Definition 5.1 (stochastically independent events) Let $(\Omega, \mathcal{P}(\Omega), \mathbb{P})$ be a discrete probability space, then events

$$A_i \subseteq \Omega, \quad i \in I \subseteq \mathbb{N}, \quad I \neq \emptyset$$

© The Author(s), under exclusive license to Springer-Verlag GmbH, DE, part of Springer Nature 2024
S. Schäffler, *Mathematics of Information*, Mathematics Study Resources 9,
https://doi.org/10.1007/978-3-662-69102-1_5

are stochastically independent, if for every non-empty finite set $J \subseteq I$ applies:

$$\mathbb{P}\left(\bigcap_{j \in J} A_j\right) = \prod_{j \in J} \mathbb{P}(A_j).$$

◁

If a random experiment is modeled by a discrete probability space $(\Omega, \mathcal{P}(\Omega), \mathbb{P})$, probability theory provides tools to make statements about the course of the underlying random experiment when the probability space is known. Mathematical statistics deals with the following problem: The random experiment to be modeled is initially described by an incomplete discrete probability space. In this description, the non-empty countable base set Ω and a set of probability measures on $\mathcal{P}(\Omega)$ are determined. The set of possible probability measures is often represented by a parameter θ from a parameter space Θ. To get at a complete mathematical description of our random experiment, we must choose a probability measure $\mathbb{P}$ from the set of possible probability measures. A key criterion of mathematical statistics is that a decision about the choice of the probability measure or about the reduction of the set of all possible probability measures depends in particular on the results of the random experiment. We thus obtain the following initial situation:

We are given a triple $(\Omega, \mathcal{P}(\Omega), \mathbb{P}_\theta)$ consisting of a non-empty countable base set Ω, a non-empty set Θ, and a set

$$\{\mathbb{P}_\theta; \theta \in \Theta\}$$

of probability measures on $\mathcal{P}(\Omega)$; furthermore, an observed result $\hat{\omega} \in \Omega$ of the random experiment with $\mathbb{P}_\theta(\{\hat{\omega}\}) > 0$ for all $\theta \in \Theta$ is given. The fact that the conditional probabilities

$$\mathbb{P}_\theta^{\{\hat{\omega}\}} : \mathcal{P}(\Omega) \to [0, 1], \quad A \mapsto \frac{\mathbb{P}_\theta(\{\hat{\omega}\} \cap A)}{\mathbb{P}_\theta(\{\hat{\omega}\})} = \begin{cases} 1 & \text{if } \hat{\omega} \in A \\ 0 & \text{if } \hat{\omega} \notin A \end{cases}, \quad \theta \in \Theta$$

do not depend on θ, we interpret as the knowledge of $\hat{\omega}$ is sufficient to make a decision about $\theta \in \Theta$. This leads us to the following definition.

Definition 5.2 (sufficient event) Let Θ be a non-empty set and let $(\Omega, \mathcal{P}(\Omega), \mathbb{P}_\theta)$ be a discrete probability space for every $\theta \in \Theta$. An event $F \in \mathcal{P}(\Omega)$ **is sufficient for θ**, if

(SE1) $\mathbb{P}_\theta(F) > 0$ for all $\theta \in \Theta$,
(SE2) $\mathbb{P}_\theta^F(A)$ does not depend on $\theta \in \Theta$ for all $A \in \mathcal{P}(\Omega)$.   ◁

According to our interpretation, a sufficient event for $\theta \in \Theta$ contains complete knowledge about $\theta \in \Theta$. The event Ω, for example, is generally not sufficient.

It can now happen that a sufficient event for $\theta \in \Theta$ is more useful than an elementary event $\{\hat{\omega}\}$, as the following example shows.

Example 5.3 (Random Generator) A random generator independently generates a sequence $b_1, \ldots, b_N$ of bits $b_i \in \{0, 1\}$, $i = 1, \ldots, N$, $N > 0$; each generated bit should be equal to one with probability $\theta \in (0, 1)$. We thus obtain $\Omega = \{0, 1\}^N$ and $\mathbb{P}_\theta$ given by

$$\mathbb{P}_\theta : \{0, 1\}^N \to [0, 1], \quad \omega = (\omega_1, \ldots, \omega_N) \mapsto \prod_{i=1}^N \theta^{\omega_i}(1 - \theta)^{1-\omega_i}.$$

Now we are interested in the events

$$F_t := \left\{ (b_1, \ldots, b_N) \in \{0, 1\}^N; \ \sum_{i=1}^N b_i = t \right\}.$$

with $t \in \{0, 1, 2, \ldots, N - 1, N\}$. For the conditional probabilities, we obtain

$$\mathbb{P}_\theta^{F_t}(\{\omega\}) = \begin{cases} 0 & \text{for all } \omega \text{ with } \sum_{i=1}^N \omega_i \neq t \\ \frac{\theta^t(1-\theta)^{(N-t)}}{\binom{N}{t}\theta^t(1-\theta)^{(N-t)}} = \frac{1}{\binom{N}{t}} & \text{for all } \omega \text{ with } \sum_{i=1}^N \omega_i = t \end{cases},$$

since

$$\prod_{i=1}^N \theta^{\omega_i}(1 - \theta)^{1-\omega_i} = \theta^{\sum_{i=1}^N \omega_i}(1 - \theta)^{\left(N - \sum_{i=1}^N \omega_i\right)}.$$

For each $t \in \{0, 1, 2, \ldots, N - 1, N\}$, the event F_t is sufficient for θ and we only need to know t about the event F_t. Hence, instead of storing an elementary event $\{\hat{\omega}_1, \ldots, \hat{\omega}_N\}$ as a result of the random experiment, which would require $\theta = \frac{1}{2} N$ bits, it is sufficient to store

$$\hat{t} = \sum_{i=1}^N \hat{\omega}_i,$$

without losing knowledge about θ. Storing $\hat{t}$ requires at most $\min\{k \in \mathbb{N}; \ k \geq \mathrm{ld}(N + 1)\}$ bits in the binary system. ◁

The example just considered suggests the following definition.

Definition 5.4 (sufficient statistic) Let Θ be a non-empty set and let $(\Omega, \mathcal{P}(\Omega), \mathbb{P}_\theta)$ be a discrete probability space for each $\theta \in \Theta$; furthermore, let $\tilde{\Omega}$ be a non-empty countable set and

$$T : \Omega \to \tilde{\Omega}$$

a mapping, then the mapping T **is called a sufficient statistic** for θ, if all events

$$\{\omega \in \Omega; \; T(\omega) = \tilde{\omega}\}, \quad \tilde{\omega} \in \tilde{\Omega},$$

are sufficient for θ.   ◁

Using theorem and definition 3.2 we know that in particular for the image measures $\mathbb{P}_{T,\theta}$ of a sufficient statistic T holds:

$$\mathbb{S}_{\mathbb{P}_\theta} \geq \mathbb{S}_{\mathbb{P}_{T,\theta}} \quad \text{for all} \quad \theta \in \Theta.$$

Furthermore, it is known that for each $\tilde{\omega} \in \tilde{\Omega}$ the conditional probability measures $\mathbb{P}_{\theta,T}^{\{\tilde{\omega}\}}$ no longer depend on θ. Therefore, it makes sense to shift decisions about $\theta \in \Theta$ to the discrete probability spaces $(\tilde{\Omega}, \mathcal{P}(\tilde{\Omega}), \mathbb{P}_{\theta,T})$, provided T represents a sufficient statistic. The decisions about $\theta \in \Theta$ itself fall into the realm of Mathematical Statistics (see for example [Held08]). By transitioning from $(\Omega, \mathcal{P}(\Omega), \mathbb{P}_\theta)$ to $(\tilde{\Omega}, \mathcal{P}(\tilde{\Omega}), \mathbb{P}_{\theta,T})$, information irrelevant for θ was thus saved.

Let's return to example 5.1 with $\Omega = \{0,1\}^N$,

$$\mathbb{P}_\theta : \{0,1\}^N \to [0,1], \quad \boldsymbol{\omega} = (\omega_1, \dots, \omega_N) \mapsto \prod_{i=1}^{N} \theta^{\omega_i}(1-\theta)^{1-\omega_i},$$

and the sufficient statistic

$$T : \Omega \to \{0, 1, \dots, (N-1), N\}, \quad \boldsymbol{\omega} \mapsto \sum_{i=1}^{N} \omega_i,$$

then we obtain the image measures $\mathbb{P}_{\theta,T}$ of T by

$$\mathbb{P}_{\theta,T}(\{k\}) = \binom{N}{k} \theta^k (1-\theta)^{N-k}, \quad k = 0, 1, \dots, (N-1), N.$$

Fig. 5.1 illustrates for $N = 50$ the entropy of $\mathbb{P}_\theta$ (solid line) and the entropy of $\mathbb{P}_{\theta,T}$ (dashed line) depending on θ.

It is clearly visible how much information can be saved without losing information about θ.

5.2   Transinformation

Starting from two non-empty sets Ω, Γ each with a finite number of elements, a probability measure

$$\mathbb{P} : \mathcal{P}(\Omega \times \Gamma) \to [0,1]$$

is given on the power set of the Cartesian product of Ω and Γ. In addition to the probability space $(\Omega \times \Gamma, \mathcal{P}(\Omega \times \Gamma), \mathbb{P})$ we obtain via projections

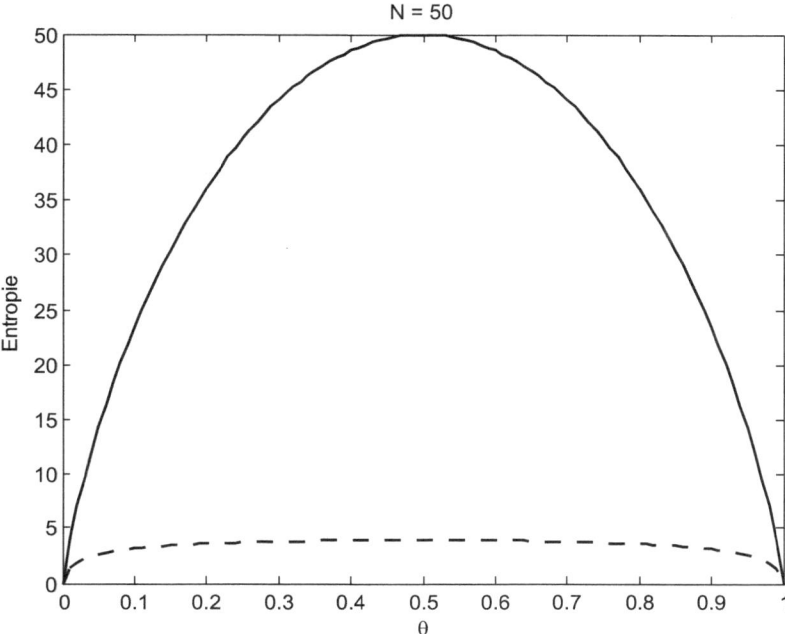

Fig. 5.1 Entropy comparison

$$P_\Omega : \quad \Omega \times \Gamma \to \Omega, \quad (\omega, \gamma) \mapsto \omega$$
$$P_\Gamma : \quad \Omega \times \Gamma \to \Gamma, \quad (\omega, \gamma) \mapsto \gamma$$

using the image measures $\mathbb{P}_{P_\Omega}, \mathbb{P}_{P_\Gamma}$ two probability spaces $(\Omega, \mathcal{P}(\Omega), \mathbb{P}_{P_\Omega})$ and $(\Gamma, \mathcal{P}(\Gamma), \mathbb{P}_{P_\Gamma})$. The question now arises as to how much information contained in random experiment $(\Omega, \mathcal{P}(\Omega), \mathbb{P}_{P_\Omega})$ one obtains when realizing the random experiment $(\Gamma, \mathcal{P}(\Gamma), \mathbb{P}_{P_\Gamma})$ and vice versa. This question leads to the concept of transinformation, which is of crucial importance in communication technology.

Theorem and Definition 5.5 (Transinformation) *Let Ω, Γ be two non-empty sets each with a finite number of elements and let there be a probability measure*

$$\mathbb{P} : \mathcal{P}(\Omega \times \Gamma) \to [0, 1]$$

on the power set of the Cartesian product of Ω and Γ. Using the projections

$$P_\Omega : \quad \Omega \times \Gamma \to \Omega, \quad (\omega, \gamma) \mapsto \omega$$
$$P_\Gamma : \quad \Omega \times \Gamma \to \Gamma, \quad (\omega, \gamma) \mapsto \gamma$$

we obtain, as is well known, the image measures

$$\mathbb{P}_{P_\Omega} : \mathcal{P}(\Omega) \to [0, 1]$$
$$\mathbb{P}_{P_\Gamma} : \mathcal{P}(\Gamma) \to [0, 1]$$

and thus two probability spaces $(\Omega, \mathcal{P}(\Omega), \mathbb{P}_{P_\Omega})$ *and* $(\Gamma, \mathcal{P}(\Gamma), \mathbb{P}_{P_\Gamma})$. *Since the sets* Ω, Γ *only have a finite number of elements, the entropies*

$$\mathbb{S}_\mathbb{P}, \; \mathbb{S}_{\mathbb{P}_{P_\Omega}} \quad and \quad \mathbb{S}_{\mathbb{P}_{P_\Gamma}}$$

are finite and we can define the **transinformation** $\mathbb{T}(\mathbb{P}_{P_\Omega}, \mathbb{P}_{P_\Gamma})$ *as follows:*

$$\mathbb{T}(\mathbb{P}_{P_\Omega}, \mathbb{P}_{P_\Gamma}) := \mathbb{S}_{\mathbb{P}_{P_\Omega}} + \mathbb{S}_{\mathbb{P}_{P_\Gamma}} - \mathbb{S}_\mathbb{P}.$$

It holds:

$$\mathbb{T}(\mathbb{P}_{P_\Omega}, \mathbb{P}_{P_\Gamma}) \geq 0$$
$$\mathbb{T}(\mathbb{P}_{P_\Omega}, \mathbb{P}_{P_\Gamma}) \leq \min\{\mathbb{S}_{\mathbb{P}_{P_\Omega}}, \mathbb{S}_{\mathbb{P}_{P_\Gamma}}\}.$$

If further

$$\mathbb{P}(\{(\omega, \gamma)\}) = \mathbb{P}_{P_\Omega}(\{\omega\}) \cdot \mathbb{P}_{P_\Gamma}(\{\gamma\}) \quad for \; all \quad \omega \in \Omega, \; \gamma \in \Gamma,$$

then follows

$$\mathbb{T}(\mathbb{P}_{P_\Omega}, \mathbb{P}_{P_\Gamma}) = 0.$$

$\triangleleft$

We assume without loss of generality that

$$\mathbb{P}_{P_\Omega}(\{\omega\}) > 0 \quad for \; all \quad \omega \in \Omega.$$

$$\mathbb{S}_\mathbb{P} = - \sum_{\substack{\omega \in \Omega \\ \gamma \in \Gamma}} \mathbb{P}(\{(\omega, \gamma)\}) \, \mathrm{ld} \, (\mathbb{P}(\{(\omega, \gamma)\}))$$

$$= - \sum_{\substack{\omega \in \Omega \\ \gamma \in \Gamma}} \mathbb{P}^{\{\omega\}}(\{\gamma\}) \mathbb{P}_{P_\Omega}(\{\omega\}) \, \mathrm{ld} \, \left(\mathbb{P}^{\{\omega\}}(\{\gamma\}) \mathbb{P}_{P_\Omega}(\{\omega\})\right)$$

$$= - \sum_{\substack{\omega \in \Omega \\ \gamma \in \Gamma}} \mathbb{P}^{\{\omega\}}(\{\gamma\}) \mathbb{P}_{P_\Omega}(\{\omega\}) \, \mathrm{ld} \, \left(\mathbb{P}_{P_\Omega}(\{\omega\})\right)$$

$$\quad - \sum_{\substack{\omega \in \Omega \\ \gamma \in \Gamma}} \mathbb{P}^{\{\omega\}}(\{\gamma\}) \mathbb{P}_{P_\Omega}(\{\omega\}) \, \mathrm{ld} \, \left(\mathbb{P}^{\{\omega\}}(\{\gamma\})\right)$$

$$= - \sum_{\omega \in \Omega} \left(\mathbb{P}_{P_\Omega}(\{\omega\}) \, \mathrm{ld} \, \left(\mathbb{P}_{P_\Omega}(\{\omega\})\right) \underbrace{\sum_{\gamma \in \Gamma} \mathbb{P}^{\{\omega\}}(\{\gamma\})}_{=1} \right)$$

$$\quad - \sum_{\substack{\omega \in \Omega \\ \gamma \in \Gamma}} \mathbb{P}^{\{\omega\}}(\{\gamma\}) \mathbb{P}_{P_\Omega}(\{\omega\}) \, \mathrm{ld} \, \left(\mathbb{P}^{\{\omega\}}(\{\gamma\})\right)$$

$$= \mathbb{S}_{\mathbb{P}_{P_\Omega}} - \sum_{\substack{\omega \in \Omega \\ \gamma \in \Gamma}} \mathbb{P}^{\{\omega\}}(\{\gamma\}) \mathbb{P}_{P_\Omega}(\{\omega\}) \, \mathrm{ld} \, \left(\mathbb{P}^{\{\omega\}}(\{\gamma\})\right).$$

Now we examine the term

$$\sum_{\omega\in\Omega}\mathbb{P}^{\{\omega\}}(\{\gamma\})\mathbb{P}_{P_\Omega}(\{\omega\})\,\mathrm{ld}\left(\mathbb{P}^{\{\omega\}}(\{\gamma\})\right)$$

in more detail. If one considers the function

$$f:[0,1]\to\mathbb{R},\quad x\mapsto\begin{cases}0 & \text{if } x=0\\ x\,\mathrm{ld}(x) & \text{if } x\neq 0\end{cases},$$

then f is strictly convex. With Jensen's inequality it follows:

$$f\left(\sum_{j=1}^{k}p_jx_j\right)\leq\sum_{j=1}^{k}p_jf(x_j),\quad x_1,\ldots,x_k\in[0,1],\ p_1,\ldots,p_k\geq 0,\ \sum_{j=1}^{k}p_j=1.$$

If we now set:

$$k=|\Omega|\quad\text{(number of elements of }\Omega\text{)},$$
$$p_j=\mathbb{P}_{P_\Omega}(\{\omega_j\}),\quad\omega_j\in\Omega,$$
$$x_j=\mathbb{P}^{\{\omega_j\}}(\{\gamma\}),\quad\omega_j\in\Omega,$$

we obtain:

$$\sum_{\omega\in\Omega}\mathbb{P}^{\{\omega\}}(\{\gamma\})\mathbb{P}_{P_\Omega}(\{\omega\})\,\mathrm{ld}\left(\mathbb{P}^{\{\omega\}}(\{\gamma\})\right)$$

$$\geq\sum_{\omega\in\Omega}\mathbb{P}^{\{\omega\}}(\{\gamma\})\mathbb{P}_{P_\Omega}(\{\omega\})\,\mathrm{ld}\left(\sum_{\omega\in\Omega}\mathbb{P}^{\{\omega\}}(\{\gamma\})\mathbb{P}_{P_\Omega}(\{\omega\})\right)$$

$$=\mathbb{P}_{P_\Gamma}(\{\gamma\})\,\mathrm{ld}\left(\mathbb{P}_{P_\Gamma}(\{\gamma\})\right).$$

Substituting above results we obtain:

$$\mathbb{S}_\mathbb{P}=\mathbb{S}_{\mathbb{P}_{P_\Omega}}-\sum_{\substack{\omega\in\Omega\\ \gamma\in\Gamma}}\mathbb{P}^{\{\omega\}}(\{\gamma\})\mathbb{P}_{P_\Omega}(\{\omega\})\,\mathrm{ld}\left(\mathbb{P}^{\{\omega\}}(\{\gamma\})\right)$$

$$\leq\mathbb{S}_{\mathbb{P}_{P_\Omega}}-\sum_{\gamma\in\Gamma}\mathbb{P}_{P_\Gamma}(\{\gamma\})\,\mathrm{ld}\left(\mathbb{P}_{P_\Gamma}(\{\gamma\})\right)$$

$$=\mathbb{S}_{\mathbb{P}_{P_\Omega}}+\mathbb{S}_{\mathbb{P}_{P_\Gamma}}$$

and thus

$$\mathbb{T}(\mathbb{P}_{P_\Omega},\mathbb{P}_{P_\Gamma})\geq 0.$$

Since due to theorem and definition 3.2

$$\mathbb{S}_{\mathbb{P}_{P_\Omega}}\leq\mathbb{S}_\mathbb{P}\quad\text{and}\quad\mathbb{S}_{\mathbb{P}_{P_\Gamma}}\leq\mathbb{S}_\mathbb{P}$$

leads to

$$\mathbb{T}(\mathbb{P}_{P_\Omega}, \mathbb{P}_{P_\Gamma}) \leq \min\{\mathbb{S}_{\mathbb{P}_{P_\Omega}}, \mathbb{S}_{\mathbb{P}_{P_\Gamma}}\}.$$

From

$$\mathbb{P}(\{(\omega, \gamma)\}) = \mathbb{P}_{P_\Omega}(\{\omega\}) \cdot \mathbb{P}_{P_\Gamma}(\{\gamma\}) \quad \text{for all} \quad \omega \in \Omega, \; \gamma \in \Gamma,$$

follows

$$\mathbb{S}_\mathbb{P} = \mathbb{S}_{\mathbb{P}_{P_\Omega}} + \mathbb{S}_{\mathbb{P}_{P_\Gamma}}$$

and thus the last assertion.

q.e.d

Example 5.6 Consider the single throw of a die and the single toss of a coin, we obtain $\Omega = \{1, 2, 3, 4, 5, 6\}$ and $\Gamma = \{K, Z\}$. Now let's assume that

$$\mathbb{P}(\{(\omega_i, K)\}) = \mathbb{P}(\{(\omega_i, Z)\}) = \frac{1}{12}, \quad i = 1, 2, \ldots, 6,$$

so

$$\mathbb{P}_{P_\Omega}(\{\omega_i\}) = \frac{1}{6}, \quad i = 1, 2, \ldots, 6,$$

$$\mathbb{P}_{P_\Gamma}(\{K\}) = \mathbb{P}_{P_\Gamma}(\{Z\}) = \frac{1}{2}$$

and thus

$$\mathbb{T}(\mathbb{P}_{P_\Omega}, \mathbb{P}_{P_\Gamma}) = \mathrm{ld}(6) + \mathrm{ld}(2) - \mathrm{ld}(12) = 0.$$

The execution of the random experiment "dice roll" in this case provides no information about the random experiment "coin toss" and vice versa.

Now we examine the random experiment "single throw of a die", but we not only record the number on top, but also the number on which the die lies. We thus obtain as a set of results

$$\Omega \times \Gamma = \{1, 2, 3, 4, 5, 6\}^2.$$

If we now again assume that each possible number on top appears with probability $\frac{1}{6}$, we obtain a probability measure on $\mathcal{P}(\{1, 2, 3, 4, 5, 6\}^2)$ given by

$$\mathbb{P}(\{(i, 7 - i)\}) = \frac{1}{6}, \quad i = 1, \ldots, 6,$$

while all other elementary events are assigned a probability of zero. It follows

$$\mathbb{T}(\mathbb{P}_{P_\Omega}, \mathbb{P}_{P_\Gamma}) = \mathrm{ld}(6) + \mathrm{ld}(6) - \mathrm{ld}(6) = \mathrm{ld}(6).$$

The transinformation is maximal because each result of the random experiment "number on top" determines the result "number on bottom" and vice versa.   ◁

In communication technology, Ω denotes a non-empty finite set consisting of symbols that a transmitter is able to send over a channel to a receiver, while the non-empty finite set Γ includes the set of symbols that can arrive at a receiver when a symbol from Ω is sent. The probability $\mathbb{P}_{P_\Omega}(\{\omega\}) > 0$, with which a symbol $\omega \in \Omega$ is sent, is known. Furthermore, the conditional probabilities

$$\mathbb{P}^{\{\omega\}}(\{\gamma\}), \quad \omega \in \Omega, \, \gamma \in \Gamma,$$

that characterize the transmission channel, are known. Thus,

$$\mathbb{P}(\{(\omega, \gamma)\}) = \mathbb{P}^{\{\omega\}}(\{\gamma\}) \cdot \mathbb{P}_{P_\Omega}(\{\omega\}) \quad \text{for all} \quad \omega \in \Omega, \, \gamma \in \Gamma$$

and

$$\mathbb{P}_{P_\Gamma}(\{\gamma\}) = \sum_{\omega \in \Omega} \mathbb{P}^{\{\omega\}}(\{\gamma\}) \cdot \mathbb{P}_{P_\Omega}(\{\omega\}) \quad \text{for all} \quad \gamma \in \Gamma$$

are known. The calculable transinformation $\mathbb{T}(\mathbb{P}_{P_\Omega}, \mathbb{P}_{P_\Gamma})$ now indicates how much information one receives on average about the sent symbol when one receives a symbol $\gamma \in \Gamma$.

If one now wants to characterize the transmission channel, on the one hand, the conditional probabilities

$$\mathbb{P}^{\{\omega\}}(\{\gamma\}), \quad \omega \in \Omega, \, \gamma \in \Gamma$$

are too complex, on the other hand, the transinformation is unsuitable because it depends on the probability measure $\mathbb{P}_{P_\Omega}$; let now $\mathbf{P}$ be the set of all probability measures defined on $\mathcal{P}(\Omega)$, which assign a probability greater than zero to each elementary event, then in communication technology for the characterization of a transmission channel, the real number

$$C := \max_{\mathbb{P}_{P_\Omega} \in \mathbf{P}} \mathbb{T}(\mathbb{P}_{P_\Omega}, \mathbb{P}_{P_\Gamma}),$$

is considered, which is referred to as **channel capacity** and which bears the designation

$$\left[\frac{\text{bit}}{\text{character to be transmitted}} \right].$$

Example 5.7 Let $\Omega = \Gamma = \{0, 1\}$ and $p_e \in [0, 1]$. Furthermore, let the error probabilities

$$\mathbb{P}^{\{0\}}(\{1\}) = \mathbb{P}^{\{1\}}(\{0\}) = p_e$$

be given, then

$$
\begin{aligned}
\mathbb{T}(\mathbb{P}_{P_\Omega}, \mathbb{P}_{P_\Gamma}) = {}& -\mathbb{P}_{P_\Omega}(\{0\})\,\mathrm{ld}(\mathbb{P}_{P_\Omega}(\{0\})) - \mathbb{P}_{P_\Omega}(\{1\})\,\mathrm{ld}(\mathbb{P}_{P_\Omega}(\{1\})) \\
& - (\mathbb{P}^{\{0\}}(\{0\}) \cdot \mathbb{P}_{P_\Omega}(\{0\}) + \mathbb{P}^{\{1\}}(\{0\}) \cdot \mathbb{P}_{P_\Omega}(\{1\})) \\
& \cdot \mathrm{ld}(\mathbb{P}^{\{0\}}(\{0\}) \cdot \mathbb{P}_{P_\Omega}(\{0\}) + \mathbb{P}^{\{1\}}(\{0\}) \cdot \mathbb{P}_{P_\Omega}(\{1\})) \\
& - (\mathbb{P}^{\{0\}}(\{1\}) \cdot \mathbb{P}_{P_\Omega}(\{0\}) + \mathbb{P}^{\{1\}}(\{1\}) \cdot \mathbb{P}_{P_\Omega}(\{1\})) \\
& \cdot \mathrm{ld}(\mathbb{P}^{\{0\}}(\{1\}) \cdot \mathbb{P}_{P_\Omega}(\{0\}) + \mathbb{P}^{\{1\}}(\{1\}) \cdot \mathbb{P}_{P_\Omega}(\{1\})) \\
& + (\mathbb{P}^{\{0\}}(\{0\}) \cdot \mathbb{P}_{P_\Omega}(\{0\}))\,\mathrm{ld}(\mathbb{P}^{\{0\}}(\{0\}) \cdot \mathbb{P}_{P_\Omega}(\{0\})) \\
& + (\mathbb{P}^{\{0\}}(\{1\}) \cdot \mathbb{P}_{P_\Omega}(\{0\}))\,\mathrm{ld}(\mathbb{P}^{\{0\}}(\{1\}) \cdot \mathbb{P}_{P_\Omega}(\{0\})) \\
& + (\mathbb{P}^{\{1\}}(\{0\}) \cdot \mathbb{P}_{P_\Omega}(\{1\}))\,\mathrm{ld}(\mathbb{P}^{\{1\}}(\{0\}) \cdot \mathbb{P}_{P_\Omega}(\{1\})) \\
& + (\mathbb{P}^{\{1\}}(\{1\}) \cdot \mathbb{P}_{P_\Omega}(\{1\}))\,\mathrm{ld}(\mathbb{P}^{\{1\}}(\{1\}) \cdot \mathbb{P}_{P_\Omega}(\{1\})) \\
= {}& -\mathbb{P}_{P_\Omega}(\{0\})\,\mathrm{ld}(\mathbb{P}_{P_\Omega}(\{0\})) - \mathbb{P}_{P_\Omega}(\{1\})\,\mathrm{ld}(\mathbb{P}_{P_\Omega}(\{1\})) \\
& - ((1 - p_e) \cdot \mathbb{P}_{P_\Omega}(\{0\}) + p_e \cdot \mathbb{P}_{P_\Omega}(\{1\})) \\
& \cdot \mathrm{ld}((1 - p_e) \cdot \mathbb{P}_{P_\Omega}(\{0\}) + p_e \cdot \mathbb{P}_{P_\Omega}(\{1\})) \\
& - (p_e \cdot \mathbb{P}_{P_\Omega}(\{0\}) + (1 - p_e) \cdot \mathbb{P}_{P_\Omega}(\{1\})) \\
& \cdot \mathrm{ld}(p_e \cdot \mathbb{P}_{P_\Omega}(\{0\}) + (1 - p_e) \cdot \mathbb{P}_{P_\Omega}(\{1\})) \\
& + ((1 - p_e) \cdot \mathbb{P}_{P_\Omega}(\{0\}))\,\mathrm{ld}((1 - p_e) \cdot \mathbb{P}_{P_\Omega}(\{0\})) \\
& + (p_e \cdot \mathbb{P}_{P_\Omega}(\{0\}))\,\mathrm{ld}(p_e \cdot \mathbb{P}_{P_\Omega}(\{0\})) \\
& + (p_e \cdot \mathbb{P}_{P_\Omega}(\{1\}))\,\mathrm{ld}(p_e \cdot \mathbb{P}_{P_\Omega}(\{1\})) \\
& + ((1 - p_e) \cdot \mathbb{P}_{P_\Omega}(\{1\}))\,\mathrm{ld}((1 - p_e) \cdot \mathbb{P}_{P_\Omega}(\{1\})) \\
= {}& (1 - p_e)\,\mathrm{ld}(1 - p_e) + p_e\,\mathrm{ld}(p_e) \\
& - ((1 - p_e)\mathbb{P}_{P_\Omega}(\{0\}) + p_e\mathbb{P}_{P_\Omega}(\{1\})) \\
& \cdot \mathrm{ld}((1 - p_e)\mathbb{P}_{P_\Omega}(\{0\}) + p_e\mathbb{P}_{P_\Omega}(\{1\})) \\
& - (p_e\mathbb{P}_{P_\Omega}(\{0\}) + (1 - p_e)\mathbb{P}_{P_\Omega}(\{1\})) \\
& \cdot \mathrm{ld}(p_e\mathbb{P}_{P_\Omega}(\{0\}) + (1 - p_e)\mathbb{P}_{P_\Omega}(\{1\})).
\end{aligned}
$$

Thus, $\mathbb{T}(\mathbb{P}_{P_\Omega}, \mathbb{P}_{P_\Gamma})$ is maximal for

$$
\mathbb{P}_{P_\Omega}(\{0\}) = \mathbb{P}_{P_\Omega}(\{1\}) = \frac{1}{2}
$$

and we obtain the channel capacity

$$
C = 1 + (1 - p_e)\,\mathrm{ld}(1 - p_e) + p_e\,\mathrm{ld}(p_e) \left[\frac{\text{bit}}{\text{character to be transmitted}} \right]
$$

For $p_e = 10^{-2}$ we obtain for example

$$
C \approx 0.9192 \left[\frac{\text{bit}}{\text{character to be transmitted}} \right].
$$

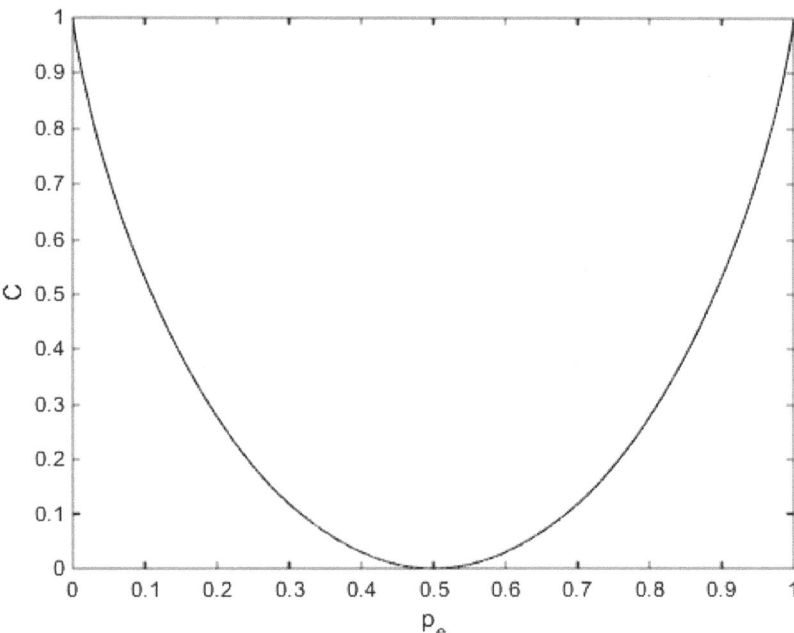

Fig. 5.2 Channel capacity

It may be surprising that for $p_e = 1$ the channel capacity is maximal; but in this case always (i.e., with probability equal to one) for the transmitted bit "1" the bit "0" is received and since always for the transmitted bit "0" the bit "1" is received, the transmitted bit can be uniquely reconstructed from the received bit. ◁

In Sect. 1.1 we learned about the Hamming code as an example of channel coding. In this case, four bits to be transmitted were always supplemented by three bits, which were chosen as the sum of a selection of the original four bits. This redundancy served to correct errors at the receiver. Now, if n is the number of bits to be transmitted and considered as a block, and k is the number of bits added to protect the n bits to be transmitted (in the Hamming code $n = 4$ and $k = 3$), then

$$R = \frac{n}{n+k}$$

is referred to as the code rate. In coding theory, the code rate is directly related to the channel capacity: Using special binary codes with a code rate of $R < C$, the probability of a transmission error using a channel of capacity C as described above can be made arbitrarily small. Conversely, if $R > C$ applies, the probability of a transmission error using a channel of capacity C as described above cannot fall below a certain positive limit (channel coding theorem by CLAUDE ELWOOD

SHANNON, see for example [Frie96]). Since the proof of this statement is not constructive, the question remains open as to how such codes with $R < C$ must be constructed. As the above example with $p_e = 10^{-2}$ shows, in this case, one would have to protect blocks of at most eleven bits with an additional bit.

Quantum Information

6

6.1   Q-Bits

The mathematical foundation of quantum information theory is formed by Hilbert spaces over the field $\mathbb{C}$ of complex numbers.

Definition 6.1 (Hilbert space over $\mathbb{C}$, Sphere) Let $\mathcal{H}$ be a vector space over $\mathbb{C}$ and let

$$\langle \cdot, \cdot \rangle_{\mathcal{H}} : \mathcal{H} \times \mathcal{H} \to \mathbb{C}$$

be a mapping (scalar product) such that for all $x, y, z \in \mathcal{H}$ and all $\lambda \in \mathbb{C}$ the following holds:

(SP1) $\langle x + y, z \rangle_{\mathcal{H}} = \langle x, z \rangle_{\mathcal{H}} + \langle y, z \rangle_{\mathcal{H}}$
(SP2) $\langle \lambda x, y \rangle_{\mathcal{H}} = \bar{\lambda} \langle x, y \rangle_{\mathcal{H}}$
          ($\bar{\lambda}$ denotes the number λ **conjugate complex**)
(SP3) $\langle x, y \rangle_{\mathcal{H}} = \overline{\langle y, x \rangle}_{\mathcal{H}}$
(SP4) $\|x\|_{\mathcal{H}}^2 := \langle x, x \rangle_{\mathcal{H}} \geq 0$ and
          $\langle x, x \rangle_{\mathcal{H}} = 0 \iff x = 0$ (neutral element of addition in $\mathcal{H}$),

then $\mathcal{H}$ **is called a Hilbert space over** $\mathbb{C}$, if for every Cauchy sequence $\{x_k\}_{k \in \mathbb{N}}$ with $x_k \in \mathcal{H}$, $k \in \mathbb{N}$, it holds: There exists a $x \in \mathcal{H}$ with

$$\lim_{k \to \infty} \|x - x_k\|_{\mathcal{H}} = 0 \quad \text{(completeness).}$$

The set

$$\mathcal{S}_{\mathcal{H}} := \{x \in \mathcal{H}; \ \|x\|_{\mathcal{H}} = 1\}$$

is referred to as the **sphere of** $\mathcal{H}$.   ◁

© The Author(s), under exclusive license to Springer-Verlag GmbH, DE, part of
Springer Nature 2024
S. Schäffler, *Mathematics of Information*, Mathematics Study Resources 9,
https://doi.org/10.1007/978-3-662-69102-1_6

Table 6.1 Key transfer with Q-bits

Sent	Basis for measurement	Classical bit after measurement
v_1	B_1	0
v_1	B_2	0 or 1 (random)
v_2	B_1	1
v_2	B_2	0 or 1 (random)
$\frac{1}{\sqrt{2}}v_1 + \frac{1}{\sqrt{2}}v_2$	B_1	0 or 1 (random)
$\frac{1}{\sqrt{2}}v_1 + \frac{1}{\sqrt{2}}v_2$	B_2	0
$\frac{1}{\sqrt{2}}v_1 - \frac{1}{\sqrt{2}}v_2$	B_1	0 or 1 (random)
$\frac{1}{\sqrt{2}}v_1 - \frac{1}{\sqrt{2}}v_2$	B_2	1

As is well known, a subset $\{v_1, \ldots, v_d\}$ of a Hilbert space $\mathcal{H}$ over $\mathbb{C}$ is a **basis** of $\mathcal{H}$, if on the one hand for $\lambda_1, \ldots, \lambda_d \in \mathbb{C}$ it holds

$$\sum_{k=1}^{d} \lambda_i v_i = 0 \quad \Longleftrightarrow \quad \lambda_1, \ldots, \lambda_d = 0$$

and if on the other hand for every $v \in \mathcal{H}$ there are unique $\hat{\lambda}_1, \ldots, \hat{\lambda}_d \in \mathbb{C}$ with

$$v = \sum_{k=1}^{d} \hat{\lambda}_i v_i.$$

If $\{v_1, \ldots, v_d\}$ is a basis of $\mathcal{H}$, then the natural number d is referred to as the **dimension** of $\mathcal{H}$; if furthermore

$$\langle v_i, v_j \rangle_{\mathcal{H}} = \begin{cases} 0 & \text{if } i \neq j \\ 1 & \text{if } i = j \end{cases}, \quad i, j \in \{1, \ldots, d\},$$

holds, one speaks of an **orthonormal basis** $\{v_1, \ldots, v_d\}$ of $\mathcal{H}$.

If $x \in B$ is combined with $|B| = 2$, then x is referred to as a **bit** in the context of classical information theory. Now, if $\mathcal{H}$ is a two-dimensional Hilbert space over $\mathbb{C}$ with an orthonormal basis $B = \{v_1, v_2\}$, then a

$$v \in \mathcal{S}_{\mathcal{H}} = \left\{ \lambda_1 v_1 + \lambda_2 v_2; \ |\lambda_1|^2 + |\lambda_2|^2 = 1, \ \lambda_1, \lambda_2 \in \mathbb{C} \right\}$$

is referred to as a **Q-bit** (short for quantum bit). From the definition of a Q-bit, it becomes clear that for a fixed $\mathcal{H}$ with orthonormal basis $\{v_1, v_2\}$ there are uncount-ably many (namely $|\mathcal{S}_{\mathcal{H}}|$) different Q-bits, while after choosing $B = \{v_1, v_2\}$ there are only two different classical bits v_1 and v_2. A function that maps a classical bit to a classical bit is given by

$$f : \{v_1, v_2\} \to \{v_1, v_2\}.$$

Functions of this kind form the basic building blocks for calculating the complexity of algorithms on classical computers (von Neumann architectures). A class of
functions f_g, which map a Q-bit to a Q-bit, is given by linear functions

$$g : \mathcal{H} \to \mathcal{H} \quad \text{such that} \quad g(\mathcal{S}_\mathcal{H}) := \{g(w); \ w \in \mathcal{S}_\mathcal{H}\} \subseteq \mathcal{S}_\mathcal{H}$$

through

$$f_g : \mathcal{S}_\mathcal{H} \to \mathcal{S}_\mathcal{H}, \quad v \mapsto g(v) \quad (\text{or briefly: } f_g = g_{|\mathcal{S}_\mathcal{H}}).$$

Functions of this kind, referred to as **gates**, now form the basic building blocks for
calculating the complexity of algorithms on quantum computers; the calculation
of a function value of f_g on a quantum computer is counted as a **one-Q-bit operation**—analogous to the calculation of a function value of f on a von Neumann
architecture, which is considered a **one-bit operation**. On a quantum computer,
the effort for a one-Q-bit operation is considered comparable to the effort for a
one-bit operation on a von Neumann architecture.

The fact that the effort to calculate the function value of a gate f_g on a von
Neumann architecture naturally requires far more than a one-bit operation is the
basis for the enormous superiority of quantum computers over von Neumann
architectures (an excellent description of the most important quantum algorithms
can be found in [StSch09]).

The linear functions described above $g : \mathcal{H} \to \mathcal{H}$ with $g(\mathcal{S}_\mathcal{H}) \subseteq \mathcal{S}_\mathcal{H}$ can be represented by unitary matrices $\mathbf{M} \in \mathbb{C}^{2,2}$ (i.e., $\overline{\mathbf{M}}^\top = \mathbf{M}^{-1}$) after choosing an orthonormal basis for $\mathcal{H}$.

A peculiarity with Q-bits lies in the process of **measurement**. In the
Copenhagen interpretation of quantum theory, a stochastic interpretation of quantum theoretical phenomena is given. Now, if a Q-bit

$$v = \hat{\lambda}_1 v_1 + \hat{\lambda}_2 v_2; \ |\hat{\lambda}_1|^2 + |\hat{\lambda}_2|^2 = 1, \ \hat{\lambda}_1, \hat{\lambda}_2 \in \mathbb{C}$$

is given with respect to the orthonormal basis $\{v_1, v_2\}$, we can interpret the non-
negative real numbers $|\hat{\lambda}_1|^2, |\hat{\lambda}_2|^2$ as probabilities; we obtain a probability space

$$(\{v_1, v_2\}, \mathcal{P}(\{v_1, v_2\}), \mathbb{P}_{v, \{v_1, v_2\}})$$

and interpret

$$\mathbb{P}_{v, \{v_1, v_2\}}(\{v_1\}) = |\hat{\lambda}_1|^2 \quad \text{as probability for } v = v_1,$$

$$\mathbb{P}_{v, \{v_1, v_2\}}(\{v_2\}) = |\hat{\lambda}_2|^2 \quad \text{as probability for } v = v_2$$

and generally with $A \in \mathcal{P}(\{v_1, v_2\})$:

$$\mathbb{P}_{v, \{v_1, v_2\}}(A) = \sum_{v_i \in A} \mathbb{P}_{v, \{v_1, v_2\}}(\{v_i\}) \quad \text{as probability for } v \in A.$$

It is important to note that two different Q-bits can imply the same probabilities,
for example

$$\frac{i}{2}v_1 + \sqrt{\frac{3}{4}}v_2 \quad \text{and} \quad \frac{1}{2}v_1 - \frac{3i}{\sqrt{12}}v_2.$$

The **measurement** of a Q-bit now means carrying out the random experiment

$$(\{v_1, v_2\}, \mathcal{P}(\{v_1, v_2\}), \mathbb{P}_{v,\{v_1,v_2\}}).$$

It is very important that the Q-bit v changes through a measurement:

If v_1 is the result of the measurement, then $v = \hat{\lambda}_1 v_1 + \hat{\lambda}_2 v_2$ changes through the measurement into the Q-bit $\frac{\hat{\lambda}_1}{|\hat{\lambda}_1|} v_1$. If v_2 is the result of the measurement, then $v = \hat{\lambda}_1 v_1 + \hat{\lambda}_2 v_2$ changes through the measurement into the Q-bit $\frac{\hat{\lambda}_2}{|\hat{\lambda}_2|} v_2$. Another measurement of this kind would always yield the result of the first measurement. The fact that a Q-bit changes through measurement is very important and has far-reaching consequences, which we will discuss later.

We can measure a Q-bit—represented with respect to the orthonormal basis $\{v_1, v_2\}$—through

$$v = \hat{\lambda}_1 v_1 + \hat{\lambda}_2 v_2; \ |\hat{\lambda}_1|^2 + |\hat{\lambda}_2|^2 = 1, \ \hat{\lambda}_1, \hat{\lambda}_2 \in \mathbb{C}$$

also with respect to another orthonormal basis $\{u_1, u_2\}$ of $\mathcal{H}$. Since there are complex numbers $\hat{\mu}_1, \hat{\mu}_2, \hat{\mu}_3, \hat{\mu}_4$ with

$$v_1 = \hat{\mu}_1 u_1 + \hat{\mu}_2 u_2 \quad \text{and} \quad v_2 = \hat{\mu}_3 u_1 + \hat{\mu}_4 u_2,$$

where

$$|\hat{\mu}_1|^2 + |\hat{\mu}_2|^2 = |\hat{\mu}_3|^2 + |\hat{\mu}_4|^2 = 1,$$

follows

$$v = (\hat{\lambda}_1 \hat{\mu}_1 + \hat{\lambda}_2 \hat{\mu}_3)u_1 + (\hat{\lambda}_1 \hat{\mu}_2 + \hat{\lambda}_2 \hat{\mu}_4)u_2.$$

We thus obtain the probability space

$$(\{u_1, u_2\}, \mathcal{P}(\{u_1, u_2\}), \mathbb{P}_{v,\{u_1,u_2\}})$$

and interpret

$$\mathbb{P}_{v,\{u_1,u_2\}}(\{u_1\}) = |\hat{\lambda}_1 \hat{\mu}_1 + \hat{\lambda}_2 \hat{\mu}_3|^2 \quad \text{as probability for } v = u_1,$$

$$\mathbb{P}_{v,\{u_1,u_2\}}(\{u_2\}) = |\hat{\lambda}_1 \hat{\mu}_2 + \hat{\lambda}_2 \hat{\mu}_4|^2 \quad \text{as probability for } v = u_2.$$

If u_1 is the result of the measurement, the Q-bit

$$v = \hat{\lambda}_1 v_1 + \hat{\lambda}_2 v_2 = (\hat{\lambda}_1 \hat{\mu}_1 + \hat{\lambda}_2 \hat{\mu}_3)u_1 + (\hat{\lambda}_1 \hat{\mu}_2 + \hat{\lambda}_2 \hat{\mu}_4)u_2$$

changes through the measurement into the Q-bit

$$\frac{(\hat{\lambda}_1 \hat{\mu}_1 + \hat{\lambda}_2 \hat{\mu}_3)}{|(\hat{\lambda}_1 \hat{\mu}_1 + \hat{\lambda}_2 \hat{\mu}_3)|} u_1.$$

If u_2 is the result of the measurement, then v changes through the measurement into the Q-bit

$$\frac{\hat{\lambda}_1\hat{\mu}_2 + \hat{\lambda}_2\hat{\mu}_4}{|\hat{\lambda}_1\hat{\mu}_2 + \hat{\lambda}_2\hat{\mu}_4|}u_2.$$

Now let's assume that the first measurement of v with respect to $\{v_1, v_2\}$ yields the result $\frac{\hat{\lambda}_2}{|\hat{\lambda}_2|}v_2$, nothing changes if we measure again with respect to $\{v_1, v_2\}$. But since

$$\frac{\hat{\lambda}_2}{|\hat{\lambda}_2|}v_2 = \frac{\hat{\lambda}_2}{|\hat{\lambda}_2|}\hat{\mu}_3 u_1 + \frac{\hat{\lambda}_2}{|\hat{\lambda}_2|}\hat{\mu}_4 u_2,$$

now a measurement with respect to $\{u_1, u_2\}$ again changes the Q-bit.

The measurement process of a Q-bit corresponds to a realization of a random experiment with two possible outcomes. Thus, the maximum entropy in a measurement of a Q-bit is given by Theorem 3.6 through

$$-\frac{1}{2}\operatorname{ld}\left(\frac{1}{2}\right) - \frac{1}{2}\operatorname{ld}\left(\frac{1}{2}\right) = 1 \text{ bit},$$

which bridges to the classical bits.

Q-bits can be used in cryptography to transmit secret keys. Let's recall the encryption from Chap. 1, where a key consisting of as many (randomly chosen) bits was necessary in the transmitter and receiver as were necessary for the message to be transmitted. Let's assume again that $B = \{v_1, v_2\}$ is an orthonormal basis of $\mathcal{H}$, then we will use four Q-bits

$$v_1, \quad v_2, \quad \frac{1}{\sqrt{2}}v_1 + \frac{1}{\sqrt{2}}v_2, \quad \frac{1}{\sqrt{2}}v_1 - \frac{1}{\sqrt{2}}v_2.$$

The transmitter and receiver agree on the assignment

$$v_1 \mathrel{\hat{=}} 0, \quad v_2 \mathrel{\hat{=}} 1, \quad \frac{1}{\sqrt{2}}v_1 + \frac{1}{\sqrt{2}}v_2 \mathrel{\hat{=}} 0, \quad \frac{1}{\sqrt{2}}v_1 - \frac{1}{\sqrt{2}}v_2 \mathrel{\hat{=}} 1.$$

The transmitter now randomly selects a sequence $\{b_k\}_{k=1,\dots,N}$, $b_k \in \{0, 1\}$, of classical bits that he wants to transmit to the receiver, and for each of these bits, he selects one of the corresponding Q-bits with probability $\frac{1}{2}$ (i.e., v_1 or $\left(\frac{1}{\sqrt{2}}v_1 + \frac{1}{\sqrt{2}}v_2\right)$ for "0", v_2 or $\left(\frac{1}{\sqrt{2}}v_1 - \frac{1}{\sqrt{2}}v_2\right)$ for "1"). Thus, we obtain a sequence $\{w_k\}_{k=1,\dots,N}$ of Q-bits with

$$w_k \in \left\{v_1, \quad v_2, \quad \frac{1}{\sqrt{2}}v_1 + \frac{1}{\sqrt{2}}v_2, \quad \frac{1}{\sqrt{2}}v_1 - \frac{1}{\sqrt{2}}v_2\right\},$$

which are transmitted (perhaps using polarized light over a fiber optic cable) to the receiver. The receiver now measures the incoming Q-bits. For this measurement, he selects one of the two orthonormal bases

$$B_1 = \{v_1, v_2\} \quad \text{oder} \quad B_2 = \left\{ \frac{1}{\sqrt{2}} v_1 + \frac{1}{\sqrt{2}} v_2, \frac{1}{\sqrt{2}} v_1 - \frac{1}{\sqrt{2}} v_2 \right\}$$

with probability $\frac{1}{2}$ for each Q-bit and performs a measurement with respect to this basis. Because of

$$v_1 = 1 \cdot v_1 + 0 \cdot v_2$$

$$v_1 = \frac{1}{\sqrt{2}} \cdot \left(\frac{1}{\sqrt{2}} v_1 + \frac{1}{\sqrt{2}} v_2 \right) + \frac{1}{\sqrt{2}} \cdot \left(\frac{1}{\sqrt{2}} v_1 - \frac{1}{\sqrt{2}} v_2 \right)$$

$$v_2 = 0 \cdot v_1 + 1 \cdot v_2$$

$$v_2 = \frac{1}{\sqrt{2}} \cdot \left(\frac{1}{\sqrt{2}} v_1 + \frac{1}{\sqrt{2}} v_2 \right) + \left(-\frac{1}{\sqrt{2}} \right) \cdot \left(\frac{1}{\sqrt{2}} v_1 - \frac{1}{\sqrt{2}} v_2 \right)$$

$$\frac{1}{\sqrt{2}} v_1 + \frac{1}{\sqrt{2}} v_2 = \frac{1}{\sqrt{2}} \cdot v_1 + \frac{1}{\sqrt{2}} \cdot v_2$$

$$\frac{1}{\sqrt{2}} v_1 + \frac{1}{\sqrt{2}} v_2 = 1 \cdot \left(\frac{1}{\sqrt{2}} \cdot v_1 + \frac{1}{\sqrt{2}} v_2 \right) + 0 \cdot \left(\frac{1}{\sqrt{2}} \cdot v_1 - \frac{1}{\sqrt{2}} v_2 \right)$$

$$\frac{1}{\sqrt{2}} v_1 - \frac{1}{\sqrt{2}} v_2 = \frac{1}{\sqrt{2}} \cdot v_1 + \left(-\frac{1}{\sqrt{2}} \right) \cdot v_2$$

$$\frac{1}{\sqrt{2}} v_1 - \frac{1}{\sqrt{2}} v_2 = 0 \cdot \left(\frac{1}{\sqrt{2}} \cdot v_1 + \frac{1}{\sqrt{2}} v_2 \right) + 1 \cdot \left(\frac{1}{\sqrt{2}} \cdot v_1 - \frac{1}{\sqrt{2}} v_2 \right)$$

the following cases can occur:

Over a classical transmission channel (e.g., telephone), the transmitter and receiver exchange information about which sent Q-bit the receiver has chosen the correct basis for (and thus the result of the measurement was not random, but has yielded the bit to be transmitted). The result of the measurement itself is not discussed. Therefore, eavesdropping on this communication provides no information about the secret key. The bits received by chance (wrong basis chosen) are not used, but ignored.

If an attacker now intercepts the transmission of the Q-bit, he can only do so by randomly choosing one of the two bases B_1 or B_2, whose use by the transmitter he knows (for example, through espionage), subjecting the corresponding Q-bit to a measurement with respect to this base, and then forwarding the measured Q-bit to the receiver. If the attacker guesses the correct base, he obtains the correct bit through the measurement of the Q-bit and the Q-bit itself is not changed by the measurement. If the attacker chooses the wrong base, the Q-bit is changed by the measurement as follows:

$$v_1 \text{ with basis } B_2 \quad \rightarrow \quad \left(\frac{1}{\sqrt{2}}v_1 + \frac{1}{\sqrt{2}}v_2\right) \quad \text{or} \quad \left(\frac{1}{\sqrt{2}}v_1 - \frac{1}{\sqrt{2}}v_2\right)$$

$$v_2 \text{ with basis } B_2 \quad \rightarrow \quad \left(\frac{1}{\sqrt{2}}v_1 + \frac{1}{\sqrt{2}}v_2\right) \quad \text{or} \quad -\left(\frac{1}{\sqrt{2}}v_1 - \frac{1}{\sqrt{2}}v_2\right)$$

$$\left(\frac{1}{\sqrt{2}}v_1 + \frac{1}{\sqrt{2}}v_2\right) \text{ with basis } B_1 \quad \rightarrow \quad v_1 \quad \text{or} \quad v_2$$

$$\left(\frac{1}{\sqrt{2}}v_1 - \frac{1}{\sqrt{2}}v_2\right) \text{ with basis } B_1 \quad \rightarrow \quad v_1 \quad \text{or} \quad -v_2.$$

If the receiver now measures the altered Q-bit and it turns out later (for example, through a phone call with the transmitter) that he has used the correct base for this measurement (otherwise the change of the Q-bit does not matter, as the resulting bit is ignored), he receives with probability $\frac{1}{2}$ a wrong bit, which he considers correct. However, unlike in classical cryptography, the transmitter and the receiver can detect that they have been eavesdropped on. To do this, bits are sent that are irrelevant for the key, so they can indeed become known. With these bits, during the classical communication between transmitter and receiver (for example, over the phone), it is not only checked whether the receiver has chosen the base matching the Q-bit, but in the case that the sent Q-bit and the chosen base match, it is also compared which bit was sent and which bit was received. If this does not match for each of these check bits, at least one Q-bit has been changed during transmission and thus intercepted. This use of Q-bits in cryptography is referred to as the BB84 protocol (see [NieChu00]).

One might now get the idea that the attacker—in order not to be discovered—simply copies the intercepted Q-bit, then performs a measurement and sends the copy to the receiver. The fact that this is not possible will be shown in Sect. 6.4.

6.2   Tensor Spaces and Multi-Q-Bits

If $x \in B$ is a classic bit, multiple bits are often combined into a symbol $y \in B^n$ (for $n = 8$ one speaks of a byte). Completely analogous to the path from the classic bit to the Q-bit, we now take the path from the symbol $y \in B^n$, $n \in \mathbb{N}$ to the Multi-Q-Bit. Since $y \in B^n$ can take exactly 2^n different values, we now need a Hilbert space (denoted by $\mathcal{H}^{\otimes n}$) of dimension $d = 2^n$, which for the special case $n = 1$ just delivers the Hilbert space $\mathcal{H}$. Important for the examination of Multi-Q-Bits are the properties of $\mathcal{H}^{\otimes n}$ preserved by isomorphism and less the following algebraic derivation. Let

$$\mathcal{M} := \{f : \mathcal{H}^n \rightarrow \mathbb{C}; \ |\{h \in \mathcal{H}^n; f(h) \neq 0\}| < \infty\}.$$

Furthermore, for $(y_1, y_2, \ldots, y_n) \in \mathcal{H}^n$ we consider the mappings

$$\delta_{(y_1,y_2,\ldots,y_n)} : \mathcal{H}^n \to \mathbb{C},$$

$$(x_1,x_2,\ldots,x_n) \mapsto \begin{cases} 1 & \text{if } (y_1,y_2,\ldots,y_n) = (x_1,x_2,\ldots,x_n) \\ 0 & \text{if } (y_1,y_2,\ldots,y_n) \neq (x_1,x_2,\ldots,x_n) \end{cases}$$

and the subspace $\mathcal{M}_0$ of $\mathcal{M}$ generated by:

$$\Big\{ \delta_{(y_1+y_1',y_2,\ldots,y_n)} - \delta_{(y_1,y_2,\ldots,y_n)} - \delta_{(y_1',y_2,\ldots,y_n)},$$

$$\delta_{(y_1,y_2+y_2',\ldots,y_n)} - \delta_{(y_1,y_2,\ldots,y_n)} - \delta_{(y_1,y_2',\ldots,y_n)},$$

$$\vdots$$

$$\delta_{(y_1,y_2,\ldots,y_n+y_n')} - \delta_{(y_1,y_2,\ldots,y_n)} - \delta_{(y_1,y_2,\ldots,y_n')},$$

$$\delta_{(ay_1,y_2,\ldots,y_n)} - a\delta_{(y_1,y_2,\ldots,y_n)},$$

$$\delta_{(y_1,ay_2,\ldots,y_n)} - a\delta_{(y_1,y_2,\ldots,y_n)},$$

$$\vdots$$

$$\delta_{(y_1,y_2,\ldots,ay_n)} - a\delta_{(y_1,y_2,\ldots,y_n)}$$

$$\text{with}$$

$$y_1,y_1',\ldots,y_n,y_n' \in \mathcal{H}, a \in \mathbb{C} \Big\}.$$

By transitioning to the quotient space we define

$$\mathcal{H}^{\otimes n} := \mathcal{M}/\mathcal{M}_0.$$

It can now be shown that the vector space $\mathcal{H}^{\otimes n}$ is isomorphic to the following vector space $\mathfrak{T}$:

Let $\{v_1, v_2\}$ be an orthonormal basis of $\mathcal{H}$, so we consider 2^n objects

$$v_{i_1} \otimes v_{i_2} \otimes \ldots \otimes v_{i_n}, \quad i_j \in \{1,2\}, \quad j = 1,\ldots,n,$$

with the following properties:

(i) For 2^n complex numbers $\lambda_{i_1,\ldots,i_n}$, $i_j \in \{1,2\}$, $j = 1,\ldots,n$, one can form a quantity $\mathbf{t}$ (a so-called **tensor**)

$$\mathbf{t} = \sum_{i_1=1}^{2} \ldots \sum_{i_n=1}^{2} \lambda_{i_1,\ldots,i_n} \cdot v_{i_1} \otimes v_{i_2} \otimes \ldots \otimes v_{i_n}.$$

.

(ii) On the set

$$\mathfrak{T} = \left\{ \sum_{i_1=1}^{2} \ldots \sum_{i_n=1}^{2} \lambda_{i_1,\ldots,i_n} \cdot v_{i_1} \otimes v_{i_2} \otimes \ldots \otimes v_{i_n}; \ \lambda_{i_1,\ldots,i_n} \in \mathbb{C} \right\}$$

of tensors, an addition

$$+ : \mathfrak{T} \times \mathfrak{T} \to \mathfrak{T}, \quad (\mathbf{t_1}, \mathbf{t_2}) \mapsto \mathbf{t_3} \quad \text{with}$$

$$\mathbf{t_1} = \sum_{i_1=1}^{2} \cdots \sum_{i_n=1}^{2} \lambda_{i_1,\ldots,i_n} \cdot v_{i_1} \otimes v_{i_2} \otimes \ldots \otimes v_{i_n}$$

$$\mathbf{t_2} = \sum_{i_1=1}^{2} \cdots \sum_{i_n=1}^{2} \mu_{i_1,\ldots,i_n} \cdot v_{i_1} \otimes v_{i_2} \otimes \ldots \otimes v_{i_n}$$

$$\mathbf{t_3} = \sum_{i_1=1}^{2} \cdots \sum_{i_n=1}^{2} (\lambda_{i_1,\ldots,i_n} + \mu_{i_1,\ldots,i_n}) \cdot v_{i_1} \otimes v_{i_2} \otimes \ldots \otimes v_{i_n}$$

and a scalar multiplication

$$\cdot : \mathbb{C} \times \mathfrak{T} \to \mathfrak{T}, \quad (z, \mathbf{t}) \mapsto \mathbf{w} \quad \text{with}$$

$$\mathbf{t} = \sum_{i_1=1}^{2} \cdots \sum_{i_n=1}^{2} \lambda_{i_1,\ldots,i_n} \cdot v_{i_1} \otimes v_{i_2} \otimes \ldots \otimes v_{i_n}$$

$$\mathbf{w} = \sum_{i_1=1}^{2} \cdots \sum_{i_n=1}^{2} (z \cdot \lambda_{i_1,\ldots,i_n}) \cdot v_{i_1} \otimes v_{i_2} \otimes \ldots \otimes v_{i_n}$$

are defined, which make $\mathfrak{T}$ into a vector space over $\mathbb{C}$.
(iii) The **fundamental identity** holds: Let

$$u_i = \lambda_{1,i} v_1 + \lambda_{2,i} v_2 \in \mathcal{H}, \quad i = 1, \ldots, n, \quad \lambda_{j,i} \in \mathbb{C}, \quad j = 1, 2,$$

then:

$$u_1 \otimes u_2 \otimes \ldots \otimes u_n = \sum_{i_1=1}^{2} \cdots \sum_{i_n=1}^{2} (\lambda_{i_1,1} \cdot \ldots \cdot \lambda_{i_n,n}) \cdot v_{i_1} \otimes v_{i_2} \otimes \ldots \otimes v_{i_n}.$$

Since the vector spaces $\mathcal{H}^{\otimes n}$ and $\mathfrak{T}$ are isomorphic, we identify $\mathcal{H}^{\otimes n}$ with $\mathfrak{T}$. The mapping

$$\langle \cdot, \cdot \rangle_{\mathcal{H}^{\otimes n}} : \mathcal{H}^{\otimes n} \times \mathcal{H}^{\otimes n} \to \mathbb{C}, \quad (\mathbf{t_1}, \mathbf{t_2}) \mapsto z \quad \text{with}$$

$$\mathbf{t_1} = \sum_{i_1=1}^{2} \cdots \sum_{i_n=1}^{2} \lambda_{i_1,\ldots,i_n} \cdot v_{i_1} \otimes v_{i_2} \otimes \ldots \otimes v_{i_n}$$

$$\mathbf{t_2} = \sum_{i_1=1}^{2} \cdots \sum_{i_n=1}^{2} \mu_{i_1,\ldots,i_n} \cdot v_{i_1} \otimes v_{i_2} \otimes \ldots \otimes v_{i_n}$$

$$z = \sum_{i_1=1}^{2} \cdots \sum_{i_n=1}^{2} \overline{\lambda_{i_1,\ldots,i_n}} \mu_{i_1,\ldots,i_n}$$

fulfills the requirements (SP1)–(SP4) of Definition 6.1 and with this scalar product, $\mathcal{H}^{\otimes n}$ becomes a Hilbert space with orthonormal basis

$$\{v_{i_1} \otimes v_{i_2} \otimes \ldots \otimes v_{i_n};\ i_j \in \{1,2\}, j = 1, \ldots, n\}.$$

With

$$\|\mathbf{t}\|^2_{\mathcal{H}^{\otimes n}} := \langle \mathbf{t}, \mathbf{t}\rangle_{\mathcal{H}^{\otimes n}} \quad \text{for all} \quad \mathbf{t} \in \mathcal{H}^{\otimes n}$$

now

$$\mathcal{S}_{\mathcal{H}^{\otimes n}} := \left\{ \mathbf{t} \in \mathcal{H}^{\otimes n};\ \|\mathbf{t}\|_{\mathcal{H}^{\otimes n}} = 1 \right\}$$

is again referred to as a **sphere of** $\mathcal{H}^{\otimes n}$ and each

$$\mathbf{s} \in \mathcal{S}_{\mathcal{H}^{\otimes n}}$$

is called a **Multi-Q-Bit** (consisting of n Q-Bits). As with Q-Bits, we again draw a class of functions F_G that map a Multi-Q-Bit to a Multi-Q-Bit; for this, we again consider linear functions

$$G : \mathcal{H}^{\otimes n} \to \mathcal{H}^{\otimes n} \quad \text{with} \quad G(\mathcal{S}_{\mathcal{H}^{\otimes n}}) \subseteq \mathcal{S}_{\mathcal{H}^{\otimes n}}$$

and establish

$$F_G = G_{|\mathcal{S}_{\mathcal{H}}^{\otimes n}}.$$

Functions of this kind are again referred to as **Gates** and can be represented by a unitary matrix $\mathbf{M} \in \mathbb{C}^{2^n, 2^n}$ depending on an orthonormal basis for $\mathcal{H}^{\otimes n}$.

For the Hilbert space $\mathcal{H}^{\otimes n}$ there are of course other orthonormal bases besides

$$\{v_{i_1} \otimes v_{i_2} \otimes \ldots \otimes v_{i_n};\ i_j \in \{1,2\}, j = 1, \ldots, n\},$$

such as

$$\frac{1}{\sqrt{2}}v_1 \otimes v_{i_2} \otimes \ldots \otimes v_{i_n} - \frac{1}{\sqrt{2}}v_2 \otimes v_{i_2} \otimes \ldots \otimes v_{i_n}, \quad i_j \in \{1,2\}, j = 2, \ldots, n,$$

$$\frac{1}{\sqrt{2}}v_1 \otimes v_{i_2} \otimes \ldots \otimes v_{i_n} + \frac{1}{\sqrt{2}}v_2 \otimes v_{i_2} \otimes \ldots \otimes v_{i_n}, \quad i_j \in \{1,2\}, j = 2, \ldots, n.$$

As with Q-Bits, the sphere $\mathcal{S}_{\mathcal{H}^{\otimes n}}$ can be represented with respect to an orthonormal basis

$$\{\mathbf{b}_1, \ldots, \mathbf{b}_{2^n}\}$$

of $\mathcal{H}^{\otimes n}$ by

$$\mathcal{S}_{\mathcal{H}^{\otimes n}} = \left\{ \sum_{i=1}^{2^n} \lambda_i \cdot \mathbf{b}_i;\ \sum_{i=1}^{2^n} |\lambda_i|^2 = 1,\ \lambda_i \in \mathbb{C} \right\}.$$

Example 6.2 Now we consider two important classes of gates for Multi-Q-bits. We start from the fixed orthonormal basis

$$\mathfrak{B}_n = \{\hat{\mathbf{b}}_1, \ldots, \hat{\mathbf{b}}_{2^n}\} := \{v_{i_1} \otimes v_{i_2} \otimes \ldots \otimes v_{i_n}; \ i_j \in \{1,2\}, j = 1, \ldots, n\}, \ n \in \mathbb{N},$$

of the Hilbert space $\mathcal{H}^{\otimes n}$ in order to be able to represent the gates as matrices. We choose the numbering

$$\hat{\mathbf{b}}_k = v_{i_1} \otimes v_{i_2} \otimes \ldots \otimes v_{i_n} \quad \text{with} \quad k = 1 + \sum_{j=1}^{n} (i_j - 1) 2^{n-j}, \quad i_j \in \{1, 2\}.$$

So it applies

$$\hat{\mathbf{b}}_1 = v_1 \otimes v_1 \otimes \ldots \otimes v_1 \otimes v_1$$

$$\hat{\mathbf{b}}_2 = v_1 \otimes v_1 \otimes \ldots \otimes v_1 \otimes v_2$$

$$\hat{\mathbf{b}}_3 = v_1 \otimes v_1 \otimes \ldots \otimes v_2 \otimes v_1$$

$$\hat{\mathbf{b}}_4 = v_1 \otimes v_1 \otimes \ldots \otimes v_2 \otimes v_2$$

$$\vdots = \vdots$$

$$\hat{\mathbf{b}}_{2^n} = v_2 \otimes v_2 \otimes \ldots \otimes v_2 \otimes v_2.$$

For two given matrices $\mathbf{A} = (a_{i,j}) \in \mathbb{C}^{p,p}$ and $\mathbf{b} = (b_{m,n}) \in \mathbb{C}^{q,q}$, $p, q \in \mathbb{N}$, the **Kronecker product** $\mathbf{A} \otimes \mathbf{b}$ is given by

$$\mathbf{A} \otimes \mathbf{b} = \begin{pmatrix} a_{1,1} \cdot \mathbf{B} & a_{1,2} \cdot \mathbf{B} & \cdots & a_{1,p} \cdot \mathbf{B} \\ a_{2,1} \cdot \mathbf{B} & a_{2,2} \cdot \mathbf{B} & \cdots & a_{2,p} \cdot \mathbf{B} \\ \vdots & \vdots & & \vdots \\ a_{p,1} \cdot \mathbf{B} & a_{p,2} \cdot \mathbf{B} & \cdots & a_{p,p} \cdot \mathbf{B} \end{pmatrix} \in \mathbb{C}^{pq,pq}.$$

For a Q-bit ($n=1$) we start with respect to $\mathfrak{B}_1$ with the gate

$$\mathbf{H}_1 := \frac{1}{\sqrt{2}} \begin{pmatrix} 1 & 1 \\ 1 & -1 \end{pmatrix}$$

and define recursively

$$\mathbf{H}_n := \mathbf{H}_1 \otimes \mathbf{H}_{n-1} \in \mathbb{R}^{2^n, 2^n}, \quad n \in \mathbb{N}.$$

For $n=2$ we get for example with respect to $\mathfrak{B}_2$:

$$\mathbf{H}_2 := \frac{1}{2} \begin{pmatrix} 1 & 1 & 1 & 1 \\ 1 & -1 & 1 & -1 \\ 1 & 1 & -1 & -1 \\ 1 & -1 & -1 & 1 \end{pmatrix}$$

The gates $\mathbf{H}_n$ are referred to as **Hadamard gates** and play an important role in quantum algorithms; we will return to this in the next section. The rows of a

Hadamard gate represent step functions, known as **Walsh functions** and are used in digital communication (see [Gauss94]).

Now we examine a Boolean function

$$h : \{0, 1\}^n \to \{0, 1\}^m, \quad \mathbf{d} \mapsto (h(\mathbf{d})_1, \ldots, h(\mathbf{d})_m), \quad n, m \in \mathbb{N}.$$

The goal is to assign an analogue to this function for Multi-Q-bits. For this purpose, we use the already known algebraic structure on $\{0, 1\}$ given by

$$0 \oplus 0 = 0, \quad 1 \oplus 0 = 0 \oplus 1 = 1, \quad 1 \oplus 1 = 0$$

and consider the function

$$\tilde{h} : \{0, 1\}^{n+m} \to \{0, 1\}^{n+m},$$

$$(\mathbf{d}, d_{n+1}, \ldots, d_{n+m}) \mapsto (\mathbf{d}, d_{n+1} \oplus h(\mathbf{d})_1, \ldots, d_{n+m} \oplus h(\mathbf{d})_m).$$

Since

$$\tilde{h}(\mathbf{d}, 0, \ldots, 0) = (\mathbf{d}, h(\mathbf{d})) \quad \text{for all} \quad \mathbf{d} \in \{0, 1\}^n,$$

represents the function h, it is bijective and has identical domain and range. In the next step, we consider the Hilbert space $\mathcal{H}^{\otimes(n+m)}$ with the basis $\mathfrak{B}_{n+m}$ and now assign through

$$K : \{0, 1\}^{n+m} \to \mathfrak{B}_{n+m}, \quad \mathbf{c} \mapsto \hat{\mathbf{b}}_r \quad \text{with} \quad r = 1 + \sum_{p=1}^{n+m} c_p 2^{p-1}$$

uniquely to each element $\mathbf{c} \in \{0, 1\}^{n+m}$ a basis element from $\mathfrak{B}_{n+m}$. Thus, we obtain the bijective mapping

$$K\tilde{h}K^{-1} : \mathfrak{B}_{n+m} \to \mathfrak{B}_{n+m}, \quad \hat{\mathbf{b}} \mapsto K(\tilde{h}(K^{-1}(\hat{\mathbf{b}}))),$$

which can be represented in the basis $\mathfrak{B}_{n+m}$ by a regular matrix

$$\mathbf{U}_h \in \{0, 1\}^{2^{n+m}, 2^{n+m}}$$

with

$$\{\mathbf{e}_1, \ldots, \mathbf{e}_{2^{n+m}}\} \to \{\mathbf{e}_1, \ldots, \mathbf{e}_{2^{n+m}}\}, \quad \mathbf{x} \mapsto \mathbf{U}_h \mathbf{x},$$

where $\{\mathbf{e}_j\}$ denotes the j-th unit vector in $\mathbb{R}^{2^{n+m}}$.

For example, if

$$h : \{0, 1\}^2 \to \{0, 1\}, \quad \mathbf{d} \mapsto d_1 \oplus d_2,$$

then

$$\tilde{h} : \{0, 1\}^3 \to \{0, 1\}^3, \quad (d_1, d_2, d_3) \mapsto (d_1, d_2, d_3 \oplus (d_1 \oplus d_2))$$

and thus

$$K\tilde{h}K^{-1} : \mathfrak{B}_3 \to \mathfrak{B}_3 \quad \text{with}$$
$$K\tilde{h}K^{-1}(\hat{\mathbf{b}}_1) = K\tilde{h}((0,0,0)) = K((0,0,0)) = \hat{\mathbf{b}}_1$$
$$K\tilde{h}K^{-1}(\hat{\mathbf{b}}_2) = K\tilde{h}((1,0,0)) = K((1,0,1)) = \hat{\mathbf{b}}_6$$
$$K\tilde{h}K^{-1}(\hat{\mathbf{b}}_3) = K\tilde{h}((0,1,0)) = K((0,1,1)) = \hat{\mathbf{b}}_7$$
$$K\tilde{h}K^{-1}(\hat{\mathbf{b}}_4) = K\tilde{h}((1,1,0)) = K((1,1,0)) = \hat{\mathbf{b}}_4$$
$$K\tilde{h}K^{-1}(\hat{\mathbf{b}}_5) = K\tilde{h}((0,0,1)) = K((0,0,1)) = \hat{\mathbf{b}}_5$$
$$K\tilde{h}K^{-1}(\hat{\mathbf{b}}_6) = K\tilde{h}((1,0,1)) = K((1,0,0)) = \hat{\mathbf{b}}_2$$
$$K\tilde{h}K^{-1}(\hat{\mathbf{b}}_7) = K\tilde{h}((0,1,1)) = K((0,1,0)) = \hat{\mathbf{b}}_3$$
$$K\tilde{h}K^{-1}(\hat{\mathbf{b}}_8) = K\tilde{h}((1,1,1)) = K((1,1,1)) = \hat{\mathbf{b}}_8.$$

The result is the symmetric matrix

$$\mathbf{U}_h = \begin{pmatrix} 1 & 0 & 0 & 0 & 0 & 0 & 0 & 0 \\ 0 & 0 & 0 & 0 & 0 & 1 & 0 & 0 \\ 0 & 0 & 0 & 0 & 0 & 0 & 1 & 0 \\ 0 & 0 & 0 & 1 & 0 & 0 & 0 & 0 \\ 0 & 0 & 0 & 0 & 1 & 0 & 0 & 0 \\ 0 & 1 & 0 & 0 & 0 & 0 & 0 & 0 \\ 0 & 0 & 1 & 0 & 0 & 0 & 0 & 0 \\ 0 & 0 & 0 & 0 & 0 & 0 & 0 & 1 \end{pmatrix}.$$

The special structure of the mapping

$$\tilde{h} : \{0,1\}^{n+m} \to \{0,1\}^{n+m},$$

namely

$$\tilde{h}(\tilde{h}(\mathbf{c})) = \mathbf{c} \quad \text{for all} \quad \mathbf{c} \in \{0,1\}^{n+m},$$

guarantees that the matrix

$$\mathbf{U}_h \in \{0,1\}^{2^{n+m},2^{n+m}}$$

is always symmetric (and thus unitary). Therefore, we are able to assign a gate

$$G_h : \mathcal{S}_{\mathcal{H}^{\otimes(n+m)}} \to \mathcal{S}_{\mathcal{H}^{\otimes(n+m)}}$$

to any Boolean function

$$h : \{0,1\}^n \to \{0,1\}^m, \quad \mathbf{d} \mapsto (h(\mathbf{d})_1, \dots h(\mathbf{d})_m), \quad n, m \in \mathbb{N}$$

that can be represented by the unitary matrix

$$\mathbf{U}_h \in \{0,1\}^{2^{n+m},2^{n+m}}$$

with respect to the basis $\mathfrak{B}_{n+m}$. The special thing is that we can also apply $\mathbf{U}_h$ to linear combinations of the basis elements of $\mathfrak{B}_{n+m}$ (represented in the basis $\mathfrak{B}_{n+m}$); we will do this in the following section.

The effort for a quantum computer to evaluate the function G_h once (i.e., a multiplication of the matrix $\mathbf{U}_h$ with a vector $\mathbf{z} \in \mathbb{C}^{2^{n+m}}$ of length one), is considered comparable to the effort of evaluating the function h on a von Neumann architecture.   ◁

6.3   Measurements

If a Multi-Q-bit

$$\mathbf{s} = \sum_{i=1}^{2^n} \hat{\lambda}_i \cdot \mathbf{b}_i, \quad \sum_{i=1}^{2^n} |\hat{\lambda}_i|^2 = 1, \ \hat{\lambda}_i \in \mathbb{C}$$

is given with respect to an orthonormal basis

$$B = \{\mathbf{b}_1, \dots, \mathbf{b}_{2^n}\}$$

of $\mathcal{H}^{\otimes n}$, we can have the non-negative real numbers $|\hat{\lambda}_i|^2$, $i = 1, \dots, 2^n$, can be interpreted again as probabilities; we obtain a probability space with $\Omega = B$ and a probability measure $\mathbb{P}_{\mathbf{s},B}$ defined on $\mathcal{P}(B)$ given by

$$\mathbb{P}_{\mathbf{s},B}(\{\mathbf{b}_i\}) = |\hat{\lambda}_i|^2, \quad i = 1, \dots, 2^n.$$

The maximum entropy is thus equal to

$$-\sum_{i=1}^{2^n} \frac{1}{2^n} \operatorname{ld}\left(\frac{1}{2^n}\right) = n \quad (\mathbf{s} \text{ consists of } n \text{ Q-Bits}),$$

when

$$|\hat{\lambda}_i|^2 = \frac{1}{2^n}, \quad i = 1 \dots, 2^n.$$

Now let $E_1, \dots, E_k, k \in \mathbb{N}$, be a partition of B, thus

(P1) $E_i \neq \emptyset$ for all $i = 1, \dots, k$,

(P2) $E_i \cap E_j = \emptyset$ for all $i \neq j$,

(P3) $\bigcup\limits_{i=1}^{k} E_i = B$,

then we obtain the probabilities

$$\mathbb{P}_{\mathbf{s},B}(E_i) = \sum_{\mathbf{b} \in E_i} \mathbb{P}_{\mathbf{s},B}(\{\mathbf{b}\}).$$

A **measurement** of the Multi-Q-bits

$$\mathbf{s} = \sum_{i=1}^{2^n} \hat{\lambda}_i \cdot \mathbf{b}_i, \quad \sum_{i=1}^{2^n} |\hat{\lambda}_i|^2 = 1, \ \hat{\lambda}_i \in \mathbb{C}$$

now consists in the execution of a random experiment with the possible results $E_1, \ldots, E_k$ and the corresponding above probabilities. The measurement of a Multi-Q-bit always refers to a partition of an orthonormal basis; the corresponding probabilities are obtained when the Multi-Q-bit to be measured is represented in the corresponding orthonormal basis. By measuring $\mathbf{s}$ the measured Multi-Q-bit is changed as follows, when $E_i = \{\mathbf{b}_{i_1}, \mathbf{b}_{i_2}, \ldots, \mathbf{b}_{i_m}\}$ represents the result of the measurement:

$$\mathbf{s} \longrightarrow \sum_{r=1}^{m} \frac{\hat{\lambda}_{i_r}}{\sqrt{\sum_{r=1}^{m} |\hat{\lambda}_{i_r}|^2}} \mathbf{b}_{i_r}.$$

For the entropy $\mathbb{S}_{\mathbb{P}_{\mathbf{s},B}}$, if we assume without loss of generality

$$\mathbb{P}_{\mathbf{s},B}(E_i) > 0 \quad \text{for all} \quad i = 1, \ldots, k :$$

$$\mathbb{S}_{\mathbb{P}_{\mathbf{s},B}} = -\sum_{\mathbf{b} \in B} \sum_{j=1}^{k} \mathbb{P}_{\mathbf{s},B}(\{\mathbf{b}\} \cap E_j) \, \text{ld} \left(\mathbb{P}_{\mathbf{s},B}(\{\mathbf{b}\} \cap E_j) \right)$$

$$= -\sum_{\mathbf{b} \in B} \sum_{j=1}^{k} \mathbb{P}_{\mathbf{s},B}^{E_j}(\{\mathbf{b}\}) \mathbb{P}_{\mathbf{s},B}(E_j) \, \text{ld} \left(\mathbb{P}_{\mathbf{s},B}^{E_j}(\{\mathbf{b}\}) \mathbb{P}_{\mathbf{s},B}(E_j) \right)$$

$$= -\sum_{\mathbf{b} \in B} \sum_{j=1}^{k} \mathbb{P}_{\mathbf{s},B}^{E_j}(\{\mathbf{b}\}) \mathbb{P}_{\mathbf{s},B}(E_j) \, \text{ld} \left(\mathbb{P}_{\mathbf{s},B}(E_j) \right)$$

$$- \sum_{\mathbf{b} \in B} \sum_{j=1}^{k} \mathbb{P}_{\mathbf{s},B}^{E_j}(\{\mathbf{b}\}) \mathbb{P}_{\mathbf{s},B}(E_j) \, \text{ld} \left(\mathbb{P}_{\mathbf{s},B}^{E_j}(\{\mathbf{b}\}) \right)$$

$$= -\sum_{j=1}^{k} \mathbb{P}_{\mathbf{s},B}(E_j) \, \text{ld} \left(\mathbb{P}_{\mathbf{s},B}(E_j) \right)$$

$$- \sum_{j=1}^{k} \left(\mathbb{P}_{\mathbf{s},B}(E_j) \sum_{\mathbf{b} \in B} \mathbb{P}_{\mathbf{s},B}^{E_j}(\{\mathbf{b}\}) \, \text{ld} \left(\mathbb{P}_{\mathbf{s},B}^{E_j}(\{\mathbf{b}\}) \right) \right)$$

$$= -\sum_{j=1}^{k} \mathbb{P}_{\mathbf{s},B}(E_j) \, \text{ld} \left(\mathbb{P}_{\mathbf{s},B}(E_j) \right)$$

$$- \sum_{j=1}^{k} \left(\mathbb{P}_{\mathbf{s},B}(E_j) \sum_{\mathbf{b} \in E_j} \mathbb{P}_{\mathbf{s},B}^{E_j}(\{\mathbf{b}\}) \, \text{ld} \left(\mathbb{P}_{\mathbf{s},B}^{E_j}(\{\mathbf{b}\}) \right) \right).$$

Thus, we can represent the entropy $\mathbb{S}_{\mathbb{P}_{s,B}}$ as the sum of the entropy of the partition

$$-\sum_{j=1}^{k} \mathbb{P}_{s,B}(E_j) \operatorname{ld}\left(\mathbb{P}_{s,B}(E_j)\right)$$

and the average conditional entropy:

$$-\sum_{j=1}^{k}\left(\mathbb{P}_{s,B}(E_j) \sum_{b \in E_j} \mathbb{P}_{s,B}^{E_j}(\{\mathbf{b}\}) \operatorname{ld}\left(\mathbb{P}_{s,B}^{E_j}(\{\mathbf{b}\})\right)\right).$$

Now let n Q-bits

$$w_i = \lambda_{1,i} v_1 + \lambda_{2,i} v_2 \quad \text{with} \quad \lambda_{1,i}, \lambda_{2,i} \in \mathbb{C}, \quad |\lambda|_{1,i}^2 + |\lambda|_{2,i}^2 = 1, \quad i = 1 \ldots, n,$$

be given, the fundamental identity results in:

$$w_1 \otimes w_2 \otimes \ldots \otimes w_n = \sum_{i_1=1}^{2} \cdots \sum_{i_n=1}^{2}(\lambda_{i_1,1} \cdot \ldots \cdot \lambda_{i_n,n}) \cdot v_{i_1} \otimes v_{i_2} \otimes \ldots \otimes v_{i_n}.$$

Now, since

$$\sum_{i_1=1}^{2} \cdots \sum_{i_n=1}^{2}|\lambda_{i_1,1} \cdot \ldots \cdot \lambda_{i_n,n}|^2 = \prod_{i=1}^{n}\left(|\lambda_{1,i}|^2 + |\lambda_{2,i}|^2\right),$$

when measuring the j-th Q-bit in $w_1 \otimes w_2 \otimes \ldots \otimes w_n$, i.e., when considering the partition

$$E_{w_j=v_1} = \{v_{i_1} \otimes v_{i_2} \otimes \ldots \otimes v_{i_n}; \ i_1, \ldots, i_n \in \{1,2\}, \ v_{i_j} = v_1\}$$
$$E_{w_j=v_2} = \{v_{i_1} \otimes v_{i_2} \otimes \ldots \otimes v_{i_n}; \ i_1, \ldots, i_n \in \{1,2\}, \ v_{i_j} = v_2\}$$

of the orthonormal basis

$$B = \{v_{i_1} \otimes v_{i_2} \otimes \ldots \otimes v_{i_n}; \ i_k \in \{1,2\}, \ k = 1, \ldots, n\},$$

the probabilities

$$\mathbb{P}_{w_1 \otimes w_2 \otimes \ldots \otimes w_n, B}(E_{w_j=v_1}) = |\lambda_{1,j}|^2$$
$$\mathbb{P}_{w_1 \otimes w_2 \otimes \ldots \otimes w_n, B}(E_{w_j=v_2}) = |\lambda_{2,j}|^2,$$

arise, which only depend on the representation of the j-th Q-bit with respect to $\{v_1, v_2\}$. If now $E_{w_j=v_1}$ is the result of the measurement, the Multi-Q-bit

$$w_1 \otimes w_2 \otimes \ldots \otimes w_n$$

transitions into the Multi-Q-bit

$$w_1 \otimes w_2 \otimes \ldots \otimes w_{j-1} \otimes \frac{\lambda_{1,j}}{|\lambda_{1,j}|} v_1 \otimes w_{j+1} \otimes \ldots \otimes w_n$$

$$= \sum_{i_1=1}^{2} \ldots \sum_{i_{j-1}=1}^{2} \sum_{i_{j+1}=1}^{2} \ldots \sum_{i_n=1}^{2} (\lambda_{i_1,1} \cdot \ldots \cdot \lambda_{i_{j-1},j-1} \cdot \frac{\lambda_{1,j}}{|\lambda_{1,j}|} \cdot \lambda_{i_{j+1},j+1} \cdot \ldots \cdot \lambda_{i_n,n})$$

$$\cdot v_{i_1} \otimes v_{i_2} \otimes \ldots \otimes v_{i_{j-1}} \otimes v_1 \otimes v_{i_{j+1}} \otimes \ldots \otimes v_{i_n}.$$

Otherwise, the Multi-Q-bit

$$w_1 \otimes w_2 \otimes \ldots \otimes w_n$$

transitions into the Multi-Q-bit

$$w_1 \otimes w_2 \otimes \ldots \otimes w_{j-1} \otimes \frac{\lambda_{2,j}}{|\lambda_{2,j}|} v_2 \otimes w_{j+1} \otimes \ldots \otimes w_n$$

$$= \sum_{i_1=1}^{2} \ldots \sum_{i_{j-1}=1}^{2} \sum_{i_{j+1}=1}^{2} \ldots \sum_{i_n=1}^{2} (\lambda_{i_1,1} \cdot \ldots \cdot \lambda_{i_{j-1},j-1} \cdot \frac{\lambda_{2,j}}{|\lambda_{2,j}|} \cdot \lambda_{i_{j+1},j+1} \cdot \ldots \cdot \lambda_{i_n,n})$$

$$\cdot v_{i_1} \otimes v_{i_2} \otimes \ldots \otimes v_{i_{j-1}} \otimes v_2 \otimes v_{i_{j+1}} \otimes \ldots \otimes v_{i_n}.$$

The Q-bits contained in

$$w_1 \otimes w_2 \otimes \ldots \otimes w_n$$

thus remain unchanged by a measurement of the j-th Q-bit given by the partition $E_{w_j=v_1}, E_{w_j=v_2}$. More interesting are Multi-Q-bits, where the corresponding Q-bits are **entangled**. If we consider, for example, the Multi-Q-bit

$$s = \frac{1}{2} v_1 \otimes v_1 + \sqrt{\frac{3}{4}} v_2 \otimes v_2 \in \mathcal{S}_{\mathcal{H}^{\otimes 2}},$$

then

$$\mathbb{P}_{s,B}(\{v_1 \otimes v_1\}) = \frac{1}{4}, \ \mathbb{P}_{s,B}(\{v_1 \otimes v_2\}) = \mathbb{P}_{s,B}(\{v_2 \otimes v_1\}) = 0, \ \mathbb{P}_{s,B}(\{v_2 \otimes v_2\}) = \frac{3}{4}.$$

applies. If we now use the partition

$$E_1 = \{v_1 \otimes v_1, v_1 \otimes v_2\} \quad \text{(first Q-Bit equal to } v_1)$$
$$E_2 = \{v_2 \otimes v_1, v_2 \otimes v_2\} \quad \text{(first Q-Bit equal to } v_2),$$

of the orthonormal basis $B = \{v_1 \otimes v_1, v_1 \otimes v_2, v_2 \otimes v_1, v_2 \otimes v_2\}$, then s is changed in the following way during a measurement:

$$s \longrightarrow v_1 \otimes v_1, \quad \text{if } E_1 \text{ is the result of a measurement,}$$
$$s \longrightarrow v_2 \otimes v_2, \quad \text{if } E_2 \text{ is the result of a measurement.}$$

Thus, if the measurement of the first Q-bit in **s** equals v_1, then this measurement also makes the second Q-bit equal to v_1 and analogously for v_2. The two Q-bits in **s** are thus entangled.

Now, the first Q-bit from **s** is sent to person A and the second Q-bit from **s** to person B. If person A now performs a measurement of his Q-bit with respect to $\{v_1, v_2\}$, the Q-bit either transitions to v_1 or to v_2. Since the two Q-bits in **s** are correspondingly entangled, the measurement of the first Q-bit from **s** at person A also transitions the second Q-bit from **s**, which is located at person B, into the corresponding state. This phenomenon forms the basis of **teleportation** of Q-bits (see NieChu00).

Now let's consider quantum algorithms; for these, a measurement is always provided as the last step. Assuming that Multi-Q-bits consisting of N Q-bits are to be manipulated by the quantum algorithm, the space forms $\mathcal{H}^{\otimes N}$ with the sphere $\mathcal{S}_{\mathcal{H}^{\otimes N}}$ and the orthonormal basis $\mathfrak{B}_N$ as the basis. A corresponding quantum algorithm—formulated with respect to the orthonormal basis $\mathfrak{B}_N$—then consists of

(1) the choice of a basis element, i.e., a unit vector $\mathbf{e} \in \mathbb{R}^{2^N}$,
(2) the sequential application of a finite number of gates $G_1, \ldots, G_p$, i.e., the calculation of the matrix multiplications

$$\mathbf{x} = \mathbf{M}_p \cdot \mathbf{M}_{p-1} \cdot \ldots \cdot \mathbf{M}_1 \mathbf{e},$$

where the unitary matrix $\mathbf{M}_j$ represents the gate G_j, $j = 1, \ldots, p$, in the basis $\mathfrak{B}_N$,

(3) a measurement of $\mathbf{x}$ with respect to a partition of $\{\mathbf{e}_1, \ldots, \mathbf{e}_{2^N}\}$.

Now let's consider an example of a quantum algorithm; for this, we start with a function

$$h : \{0, 1\}^n \rightarrow \{0, 1\}, \quad n \in \mathbb{N},$$

which we know is either constant (i.e., always delivers the same function value) or balanced (i.e., for 2^{n-1} arguments it delivers the function value 0 and for 2^{n-1} arguments it delivers the function value 1). The goal is to find an algorithm that determines which of the two possible properties h has. On a von Neumann architecture, one would simply insert the points of the domain; as soon as one has two different function values, the question is decided; otherwise, the question is decided when one has $(2^{n-1} + 1)$ identical function values. Thus, one needs at least two and at most $(2^{n-1} + 1)$ evaluations of the function h. Now we examine a corresponding quantum algorithm to answer this question (the algorithm of DEUTSCH/JOSZA, see StSch09). Obviously, $N = n + 1$ and we need the gate G_h associated with h, given by a unitary matrix

$$\mathbf{U}_h \in \{0, 1\}^{2^N, 2^N},$$

the Hadamard gate $\mathbf{H}_N$ and the partition

$$E_1 = \{\mathbf{e}_{2^n+1}\} \quad \text{and} \quad E_2 = \{\mathbf{e}_1, \ldots, \mathbf{e}_{2^N}\} \setminus E_1.$$

The corresponding quantum algorithm is then:

(1) Select $\mathbf{e}_{2^n+1}$.
(2.1) Calculate

$$\mathbf{x}_1 = \mathbf{H}_N \cdot \mathbf{e}_{2^n+1}.$$

(2.2) Calculate

$$\mathbf{x}_2 = \mathbf{U}_h \cdot \mathbf{x}_1.$$

(2.3) Calculate

$$\mathbf{x}_3 = \mathbf{H}_N \cdot \mathbf{x}_2.$$

(3) Perform a measurement of $\mathbf{x}_3$ with respect to the partition

$$E_1 = \{\mathbf{e}_{2^n+1}\} \quad \text{and} \quad E_2 = \{\mathbf{e}_1, \ldots, \mathbf{e}_{2^N}\} \setminus E_1.$$

Due to the definition of the Hadamard gates,

$$\mathbf{x}_1 = \frac{1}{\sqrt{2^{n+1}}} \begin{pmatrix} 1 \\ \vdots \\ 1 \\ -1 \\ \vdots \\ -1 \end{pmatrix}$$

holds with 2^n positive and 2^n negative entries. In terms of probability theory, this corresponds to a Multi-Q-bit with maximum entropy equal to N. For $\mathbf{x}_2$ and $\mathbf{x}_3$ three cases are now to be distinguished:

Case 1: The function h is constant with a function value equal to zero.
In this case, $\mathbf{x}_1 = \mathbf{x}_2$ and thus the $(2^n + 1)$-th component of $\mathbf{x}_3$ is equal to 1.

Case 2: The function h is constant with a function value equal to one.
In this case, $\mathbf{x}_2 = -\mathbf{x}_1$ and thus the $(2^n + 1)$-th component of $\mathbf{x}_3$ is equal to -1.

Case 3: The function h is balanced.
In this case, half of the first 2^n components of $\mathbf{x}_2$ are equal to $\frac{1}{\sqrt{2^{n+1}}}$ and half of the first 2^n components of $\mathbf{x}_2$ are equal to $-\frac{1}{\sqrt{2^{n+1}}}$. Furthermore, half of the second 2^n components of $\mathbf{x}_2$ are equal to $\frac{1}{\sqrt{2^{n+1}}}$ and half of the second 2^n components of $\mathbf{x}_2$ are equal to $-\frac{1}{\sqrt{2^{n+1}}}$. Thus, the $(2^n + 1)$-th component of $\mathbf{x}_3$ is equal to 0.

If the measurement of the Multi-Q-bit $\mathbf{x}_3$ yields the result E_1, then the function h is constant (with probability 1); if the measurement of the Multi-Q-bit $\mathbf{x}_3$ the result E_2, then the function h is balanced with probability 1. Therefore, only one function evaluation on a quantum computer is necessary for the decision.

If one applies a unit vector to a Hadamard gate, one always obtains a uniform distribution, i.e., a Multi-Q-bit with maximum entropy. The above algorithm works because a special uniform distribution (by choosing the unit vector $\mathbf{e}_{2^n+1}$) is chosen; thus, not only the distribution is important, but also its representation. Therefore, it is not readily possible to map a quantum algorithm efficiently onto a von Neumann architecture using stochastic algorithms.

6.4   Copying

If one has a classical bit $b \in \{0, 1\}$ given, this bit can be copied. For this, one considers, for example, a function

$$f : \{0, 1\}^2 \to \{0, 1\}^2, \quad (b_1, b_2) \mapsto (b_1, b_1 \oplus b_2)$$

(reminder: $0 \oplus 1 = 1 \oplus 0 = 1, 0 \oplus 0 = 1 \oplus 1 = 0$). It holds:

$$f(b, 0) = (b, b) \quad \text{for all} \quad b \in \{0, 1\}.$$

Now we are looking for a gate
$$F : \mathcal{S}_{\mathcal{H}^{\otimes 2}} \to \mathcal{S}_{\mathcal{H}^{\otimes 2}}$$

such that an $\hat{w} \in \mathcal{S}_{\mathcal{H}}$ exists with

$$F(\psi \otimes \hat{w}) = \psi \otimes \psi \quad \text{for all} \quad \psi \in \mathcal{S}_{\mathcal{H}}.$$

The Q-bit $\hat{w}$ corresponds to an empty "paper" onto which ψ is to be copied. Now we choose an orthonormal basis $\{v_1, v_2\}$ of $\mathcal{H}$, an orthonormal basis

$$\{v_1 \otimes v_1, v_2 \otimes v_1, v_1 \otimes v_2, v_2 \otimes v_2\}$$

of $\mathcal{H}^{\otimes 2}$ as well as

$$\psi_1 = \hat{\lambda}_1 v_1 + \hat{\lambda}_2 v_2, \quad |\hat{\lambda}_1|^2 + |\hat{\lambda}_2|^2 = 1$$
$$\psi_2 = \hat{\mu}_1 v_1 + \hat{\mu}_2 v_2, \quad |\hat{\mu}_1|^2 + |\hat{\mu}_2|^2 = 1$$
$$\hat{w} = \hat{\sigma}_1 v_1 + \hat{\sigma}_2 v_2, \quad |\hat{\sigma}_1|^2 + |\hat{\sigma}_2|^2 = 1,$$

so from

$$F(\psi_1 \otimes \hat{w}) = \psi_1 \otimes \psi_1$$
$$F(\psi_2 \otimes \hat{w}) = \psi_2 \otimes \psi_2$$

follows the existence of a unitary matrix $\mathbf{M} \in \mathbb{C}^{4,4}$ with

$$\mathbf{M} \begin{pmatrix} \hat{\lambda}_1 \hat{\sigma}_1 \\ \hat{\lambda}_2 \hat{\sigma}_1 \\ \hat{\lambda}_1 \hat{\sigma}_2 \\ \hat{\lambda}_2 \hat{\sigma}_2 \end{pmatrix} = \begin{pmatrix} \hat{\lambda}_1 \hat{\lambda}_1 \\ \hat{\lambda}_2 \hat{\lambda}_1 \\ \hat{\lambda}_1 \hat{\lambda}_2 \\ \hat{\lambda}_2 \hat{\lambda}_2 \end{pmatrix}$$

$$\mathbf{M} \begin{pmatrix} \hat{\mu}_1 \hat{\sigma}_1 \\ \hat{\mu}_2 \hat{\sigma}_1 \\ \hat{\mu}_1 \hat{\sigma}_2 \\ \hat{\mu}_2 \hat{\sigma}_2 \end{pmatrix} = \begin{pmatrix} \hat{\mu}_1 \hat{\mu}_1 \\ \hat{\mu}_2 \hat{\mu}_1 \\ \hat{\mu}_1 \hat{\mu}_2 \\ \hat{\mu}_2 \hat{\mu}_2 \end{pmatrix} .$$

Now, if we form the scalar product of the two vectors to the left of the equals sign and the scalar product of the two vectors to the right of the equals sign, we obtain

$$\left\langle \mathbf{M} \begin{pmatrix} \hat{\lambda}_1 \hat{\sigma}_1 \\ \hat{\lambda}_2 \hat{\sigma}_1 \\ \hat{\lambda}_1 \hat{\sigma}_2 \\ \hat{\lambda}_2 \hat{\sigma}_2 \end{pmatrix} , \mathbf{M} \begin{pmatrix} \hat{\mu}_1 \hat{\sigma}_1 \\ \hat{\mu}_2 \hat{\sigma}_1 \\ \hat{\mu}_1 \hat{\sigma}_2 \\ \hat{\mu}_2 \hat{\sigma}_2 \end{pmatrix} \right\rangle_{\mathcal{H}^{\otimes 2}} = \left\langle \begin{pmatrix} \hat{\lambda}_1 \hat{\sigma}_1 \\ \hat{\lambda}_2 \hat{\sigma}_1 \\ \hat{\lambda}_1 \hat{\sigma}_2 \\ \hat{\lambda}_2 \hat{\sigma}_2 \end{pmatrix} , \begin{pmatrix} \hat{\mu}_1 \hat{\sigma}_1 \\ \hat{\mu}_2 \hat{\sigma}_1 \\ \hat{\mu}_1 \hat{\sigma}_2 \\ \hat{\mu}_2 \hat{\sigma}_2 \end{pmatrix} \right\rangle_{\mathcal{H}^{\otimes 2}} =$$

$$= \sum_{i=1}^{2} \sum_{j=1}^{2} \overline{\hat{\lambda}_i \hat{\sigma}_j} \hat{\mu}_i \hat{\sigma}_j = \sum_{i=1}^{2} \sum_{j=1}^{2} \overline{\hat{\lambda}_i} \hat{\mu}_i |\hat{\sigma}_j|^2 = \sum_{i=1}^{2} \overline{\hat{\lambda}_i} \hat{\mu}_i =$$

$$= \langle \psi_1, \psi_2 \rangle_{\mathcal{H}}$$

and

$$\left\langle \begin{pmatrix} \hat{\lambda}_1 \hat{\lambda}_1 \\ \hat{\lambda}_2 \hat{\lambda}_1 \\ \hat{\lambda}_1 \hat{\lambda}_2 \\ \hat{\lambda}_2 \hat{\lambda}_2 \end{pmatrix} , \begin{pmatrix} \hat{\mu}_1 \hat{\mu}_1 \\ \hat{\mu}_2 \hat{\mu}_1 \\ \hat{\mu}_1 \hat{\mu}_2 \\ \hat{\mu}_2 \hat{\mu}_2 \end{pmatrix} \right\rangle_{\mathcal{H}^{\otimes 2}} = \sum_{i=1}^{2} \sum_{j=1}^{2} \overline{\hat{\lambda}_i \hat{\lambda}_j} \hat{\mu}_i \hat{\mu}_j = \left(\overline{\hat{\lambda}_1} \hat{\mu}_1 + \overline{\hat{\lambda}_2} \hat{\mu}_2 \right)^2 =$$

$$= \langle \psi_1, \psi_2 \rangle_{\mathcal{H}}^2 .$$

Thus, we obtain the requirement for ψ_1, ψ_2:

$$\langle \psi_1, \psi_2 \rangle_{\mathcal{H}} = \langle \psi_1, \psi_2 \rangle_{\mathcal{H}}^2, \quad \text{hence} \quad \langle \psi_1, \psi_2 \rangle_{\mathcal{H}} = 0 \text{ or } \langle \psi_1, \psi_2 \rangle_{\mathcal{H}} = 1.$$

Therefore, copying a Q-bit in the form described above is not possible, since

$$F(\psi_1 \otimes \hat{w}) = \psi_1 \otimes \psi_1$$
$$F(\psi_2 \otimes \hat{w}) = \psi_2 \otimes \psi_2$$

would have to be possible for all $\psi_1, \psi_2 \in \mathcal{S}_{\mathcal{H}}$.

Part III
General Systems

The Entropy of Partitions

7

7.1 Uncountable Outcomes

When considering a discrete probability space $(\Omega, \mathcal{P}(\Omega), \mathbb{P})$, the following properties were required for the probability measure $\mathbb{P}$:

(P1) $\mathbb{P} : \mathcal{P}(\Omega) \to [0, 1]$, where $\mathcal{P}(\Omega)$ denotes the power set of Ω.

(P2) $\mathbb{P}(\emptyset) = 0$, $\mathbb{P}(\Omega) = 1$.

(P3) For every sequence $\{A_i\}_{i \in \mathbb{N}}$ of pairwise disjoint sets with $A_i \in \mathcal{P}(\Omega)$, $i \in \mathbb{N}$, the following holds:

$$\mathbb{P}\left(\bigcup_{i=1}^{\infty} A_i\right) = \sum_{i=1}^{\infty} \mathbb{P}(A_i).$$

A non-empty set Ω is called **uncountable**, if there is no surjective mapping $\mathbb{N} \to \Omega$ (in symbols: $|\Omega| > |\mathbb{N}|$). It would now be obvious to demand the properties (P1)–(P3) for a probability measure $\mathbb{P}$ even when there are uncountably many outcomes in Ω. Unfortunately, it turns out that for uncountable Ω there are no mappings $\mathbb{P}$ of this kind that are useful in practice (see, for example, [Wagon85]); the uncountability of Ω severely limits the possibilities of finding a $\mathbb{P}$ with the properties (P1)–(P3). Since on the one hand we cannot do without probability spaces with uncountable outcome sets, and on the other hand the properties specified by (P1)–(P3) are fundamentally indispensable, within the framework of measure theory we have moved to defining the domain of $\mathbb{P}$ (hereinafter referred to as $\mathcal{D}$ $(\subseteq \mathcal{P}(\Omega))$ to limit (i.e., not to demand the power set of Ω anymore), thus expanding the possibilities for the choice of $\mathbb{P}$; otherwise, the properties (P1)–(P3) should apply to $\mathcal{D}$ instead of $\mathcal{P}(\Omega)$. This immediately implies that $\mathcal{D}$ must exhibit a certain minimal structure:

S. Schäffler, *Mathematics of Information*, Mathematics Study Resources 9,
https://doi.org/10.1007/978-3-662-69102-1_7

(i) $\Omega, \emptyset \in \mathcal{D}$ due to (P2).
(ii) For each sequence $\{A_i\}_{i \in \mathbb{N}}$ of pairwise disjoint sets with $A_i \in \mathcal{D}, i \in \mathbb{N}$, the following applies:

$$\bigcup_{i=1}^{\infty} A_i \in \mathcal{D} \quad \text{because of (P3).}$$

Since on the one hand, one is dependent on finding as many subsets of Ω in $\mathcal{D}$ as possible, since only these sets can be assigned a probability, and since on the other hand, one should not choose $\mathcal{D}$ too extensively, as otherwise the existence of practically relevant probability measures is endangered, one wishes for $\mathcal{D}$ in addition to (i) and (ii) another important structural feature: If one chooses an (unstructured) set $\mathcal{M} \subset \mathcal{P}(\Omega)$ (subsets of Ω, which one absolutely wants to assign a probability), there should be a **smallest** set $\mathcal{D} \subseteq \mathcal{P}(\Omega)$ that fulfills (i) and (ii) and for which $\mathcal{M} \subseteq \mathcal{D}$ applies; in other words: If $\mathcal{D}_1 \subseteq \mathcal{P}(\Omega)$ and $\mathcal{D}_2 \subseteq \mathcal{P}(\Omega)$ are two sets that fulfill (i) and (ii), then also $\mathcal{D}_1 \cap \mathcal{D}_2$ these two properties fulfill, because then there would be the smallest set

$$\mathcal{D}(\mathcal{M}) := \bigcap_{\mathcal{D} \in \mathbf{D}} \mathcal{D},$$

that fulfills (i) and (ii) and that contains the set $\mathcal{M}$, where

$$\mathbf{D} = \{\mathcal{G} \subseteq \mathcal{P}(\Omega); \ \mathcal{M} \subseteq \mathcal{G} \text{ and } \mathcal{G} \text{ fulfills (i) and (ii)}\}.$$

These requirements lead to the structural features of a σ-field over Ω.

Definition 7.1 (σ-field) Let Ω be a non-empty set. A set $\mathcal{S} \subseteq \mathcal{P}(\Omega)$ is called a σ**-field** over Ω, if the following axioms are fulfilled:

(S1) $\Omega \in \mathcal{S}$.
(S2) From $A \in \mathcal{S}$ follows $A^c := \{\omega \in \Omega; \ \omega \notin A\} \in \mathcal{S}$.
(S3) From $A_i \in \mathcal{S}, i \in \mathbb{N}$, follows $\bigcup_{i=1}^{\infty} A_i \in \mathcal{S}$. ◁

The great advantage in the structural features of a σ-field over Ω lies not only in the compatibility with the requirements for the mapping $\mathbb{P}$ (i.e., properties (i) and (ii)), but in the fact that the intersection of two σ-fields over Ω is again a σ-field over Ω. If you now have a wish list $\mathcal{M}$ of subsets of Ω, to which you definitely want to assign a probability, then with

$$\sigma(\mathcal{M}) := \bigcap_{\mathcal{F} \in \Sigma} \mathcal{F}$$

the smallest σ-field over Ω is given, which contains $\mathcal{M}$, where Σ represents the set of all σ-fields over Ω that contain $\mathcal{M}$.

In summary, a probability space is given by the outcome set Ω, a σ-algebra $\mathcal{S}$ over Ω and a probability measure $\mathbb{P}$, thus a mapping $\mathbb{P}$ defined on $\mathcal{S}$, which fulfills the conditions

(P1') $\quad \mathbb{P} : \mathcal{S} \to [0, 1]$

(P2) $\quad \mathbb{P}(\emptyset) = 0, \mathbb{P}(\Omega) = 1$

(P3') $\quad$ For each sequence $\{A_i\}_{i \in \mathbb{N}}$ of pairwise disjoint sets with $A_i \in \mathcal{S}, i \in \mathbb{N}$, applies:

$$\mathbb{P}\left(\bigcup_{i=1}^{\infty} A_i\right) = \sum_{i=1}^{\infty} \mathbb{P}(A_i).$$

In the case of $\Omega = \mathbb{R}^n$, $n \in \mathbb{N}$, the choice

$$\mathcal{M} = \{A \subseteq \mathbb{R}^n; \ A \text{ open}\}$$

has proven to be effective. The σ-field

$$\mathcal{B}^n := \sigma(\mathcal{M})$$

is called Borel σ-algebra over $\mathbb{R}^n$. Although

$$\mathcal{B}^n \neq \mathcal{P}(\mathbb{R}^n),$$

are in $\mathcal{B}^n$ all relevant subsets of the $\mathbb{R}^n$ (including the closed and compact subsets) included. Furthermore, for all practically relevant questions, there are suitable probability measures defined on $\mathcal{B}^n$. If Ω is countable, as before, $\mathcal{S} = \mathcal{P}(\Omega)$ can always be chosen (obviously, $\mathcal{P}(\Omega)$ is always a σ-field over Ω). A tuple $(\Omega, \mathcal{S})$ consisting of a non-empty result set Ω and a σ-field $\mathcal{S}$ over Ω is referred to as a **measure space**. The elements of the σ-field $\mathcal{S}$ are called **events**.

In general, for every uncountable result set Ω, a "suitable" σ-field $\mathcal{S}$ must be chosen. In the chapter on stationary information sources, we will carry out this process in detail for the set

$$\Omega = A^{\mathbb{Z}} := \{f : \mathbb{Z} \to A\},$$

where A represents a non-empty set with a finite number of elements (the so-called "character set"). In contrast to discrete probability spaces, we can no longer assume that the elementary events are elements of the σ-fields for σ-fields over uncountable sets Ω. If this is the case, a probability measure

$$\mathbb{P} : \mathcal{S} \to [0, 1]$$

can no longer be determined by specifying the probabilities for the elementary events, as this would require the summation of uncountably many summands. Therefore, the concept of entropy cannot be directly adopted from the theory of discrete probability spaces, but must be adapted in the following section.

7.2 Entropy

In a discrete probability space $(\Omega, \mathcal{P}(\Omega), \mathbb{P})$ the entropy was given by

$$\mathbb{S}_{\mathbb{P}} = -\sum_{\omega \in \Omega} \mathbb{P}(\{\omega\}) \, \mathrm{ld}(\mathbb{P}(\{\omega\})).$$

The events considered in the calculation of entropy $\{\omega\}$, $\omega \in \Omega$, thus fulfill the following properties:

(i)   All events are not empty.
(ii)   The intersection of two different events is always the empty set.
(iii)   The union of all events results in the outcome set Ω.

Let now $I \subseteq \mathbb{N}$ and $\{E_i \subseteq \Omega; \ i \in I\}$ be a set of events for which (i)–(iii) apply (thus a **partition** of Ω), we obtain:

$$
\begin{aligned}
\mathbb{S}_\mathbb{P} &= -\sum_{\omega \in \Omega} \mathbb{P}(\{\omega\}) \operatorname{ld}(\mathbb{P}(\{\omega\})) = -\sum_{i \in I} \sum_{\omega \in E_i} \mathbb{P}(\{\omega\}) \operatorname{ld}(\mathbb{P}(\{\omega\})) \\
&\geq -\sum_{i \in I} \sum_{\omega \in E_i} \mathbb{P}(\{\omega\}) \operatorname{ld}(\mathbb{P}(E_i)) = -\sum_{i \in I} \operatorname{ld}(\mathbb{P}(E_i)) \sum_{\omega \in E_i} \mathbb{P}(\{\omega\}) \\
&= -\sum_{i \in I} \mathbb{P}(E_i) \operatorname{ld}(\mathbb{P}(E_i)).
\end{aligned}
$$

Thus, in defining entropy, we have chosen from all possible partitions of Ω the one that maximizes the average amount of information. We now continue this approach in defining entropy for general probability spaces (including those with uncountably many outcomes).

Definition 7.2 (Partition of events) Let $(\Omega, \mathcal{S}, \mathbb{P})$ be a probability space and I a non-empty set with $|I| \leq |\mathbb{N}|$ (thus I with finitely many or at most countably infinite elements), then a set

$$
P_\mathcal{S} = \{E_i \in \mathcal{S}; \ i \in I\}
$$

is called a **partition of Ω of events**, if the following applies:

(i) $E_i \neq \emptyset$ for all $i \in I$,
(ii) $E_i \cap E_j = \emptyset$ for all $i, j \in I, i \neq j$,
(iii) $\bigcup_{i \in I} E_i = \Omega.\triangleleft$

The following definition is now obvious.

Definition 7.3 ((Shannon-)Entropy) Let $(\Omega, \mathcal{S}, \mathbb{P})$ be a probability space and $\Pi_\mathcal{S}$ the set of all partitions of Ω from events, then the size

$$
\mathbb{S}_\mathbb{P} := \sup_{P_\mathcal{S} \in \Pi_\mathcal{S}} \left\{ -\sum_{E \in P_\mathcal{S}} \mathbb{P}(E) \operatorname{ld}(\mathbb{P}(E)) \right\}
$$

is referred to as **entropy** or **Shannon entropy** of $(\Omega, \mathcal{S}, \mathbb{P})$. $\triangleleft$
    Obviously,

$$
0 \leq \mathbb{S}_\mathbb{P} \leq \infty.
$$

holds. In probability theory, based on a probability space $(\Omega, \mathcal{S}, \mathbb{P})$ and a measure space $(\Omega', \mathcal{S}')$, **random variable**

$$\mathbf{X} : \Omega \to \Omega',$$

are considered such that:

$$\mathbf{X}^{-1}(A') \in \mathcal{S} \quad \text{for all} \quad A' \in \mathcal{S}'.$$

This property is referred to as $\mathcal{S}$-$\mathcal{S}'$-**measurability** of $\mathbf{X}$. In measure theory, there are various criteria to prove the measurability of a mapping (see for example [Bau92]). We will often omit these concrete proofs in the following, as they unnecessarily interrupt the main line of thought. However, it should be clear that this would need to be proven for every mapping declared as measurable in the following.

A random variable serves to highlight certain aspects of a random experiment given by $(\Omega, \mathcal{S}, \mathbb{P})$ and to hide unimportant aspects.

Example 7.4 A random generator produces a real number in the interval $[0, 1]$. Since $\Omega = [0, 1]$ is uncountable, we choose the σ-field

$$\mathcal{S} = \{[0, 1] \cap A; \ A \in \mathcal{B}\}$$

over $[0, 1]$, where $\mathcal{B}$ represents the Borel σ-field over $\mathbb{R}$. From measure theory, it is known that a probability measure $\mathbb{P}$ on $\mathcal{S}$ is determined by specifying the probabilities

$$\mathbb{P}((a, b]) := b - a, \quad 0 \le a < b \le 1.$$

From the properties (P1'), (P2) and (P3') it follows:

$$\mathbb{P}([a, b]) = b - a, \quad 0 \le a \le b \le 1$$

and thus

$$\mathbb{P}(\{x\}) = 0 \quad \text{for all} \quad x \in [0, 1].$$

For each $N \in \mathbb{N}$,

$$\{0\}, \left(0, \frac{1}{2^N}\right], \left(\frac{1}{2^N}, \frac{2}{2^N}\right], \dots, \left(\frac{2^N - 1}{2^N}, 1\right]$$

is a partition of $[0, 1]$ from events and

$$\mathbb{S}_{\mathbb{P}} \ge -\mathbb{P}(\{0\}) \operatorname{ld}(\mathbb{P}(\{0\})) - \sum_{i=1}^{2^N} \mathbb{P}\left(\left(\frac{i-1}{2^N}, \frac{i}{2^N}\right]\right) \operatorname{ld}\left(\mathbb{P}\left(\left(\frac{i-1}{2^N}, \frac{i}{2^N}\right]\right)\right) = N.$$

applies. Thus,

$$\mathbb{S}_{\mathbb{P}} = \infty.$$

applies. Now we consider the mapping

$$\mathbf{X} : [0, 1] \to \{1, 2, \dots, 10\}, \quad x \mapsto \max_{k \in \{1, 2, \dots, 10\}} \left\{k; \ \frac{1}{k} \ge x\right\}.$$

Since

$$\mathbf{X}^{-1}(\{k\}) = \begin{cases} \left(\frac{1}{k+1}, \frac{1}{k}\right] & \text{if } 1 \leq k \leq 9 \\ \left[0, \frac{1}{10}\right] & \text{if } k = 10 \end{cases},$$

$\mathbf{X}$ is a random variable and it holds:

$$\mathbb{S}_{\mathbb{P}_{\mathbf{X}}} = -\frac{1}{10} \, \text{ld}\left(\frac{1}{10}\right) - \sum_{k=1}^{9} \left(\frac{1}{k} - \frac{1}{k+1}\right) \text{ld}\left(\frac{1}{k} - \frac{1}{k+1}\right) \approx 2.33 \text{ bit.}$$

The random variable $\mathbf{X}$ thus highlights the aspect "The random number lies in one of the intervals $\left(0, \frac{1}{10}\right], \left(\frac{1}{10}, \frac{1}{9}\right], \ldots, \left(\frac{1}{2}, 1\right]$" and fades out everything else. Intuitively, one expects that when transitioning from $([0, 1], \mathcal{S}, \mathbb{P})$ to $(\{1, 2, \ldots, 10\}, \mathcal{P}(\{1, 2, \ldots, 10\}), \mathbb{P}_{\mathbf{X}})$ no additional information is gained, but rather information is lost, which is also confirmed by the calculation.◁

The phenomenon observed in the above example is general.

▶ **Theorem 7.5 (Entropy and Random Variable)** *Let* $(\Omega, \mathcal{S}, \mathbb{P})$ *be a probability space,* $(\Omega', \mathcal{S}')$ *a measure space,*

$$\mathbf{X} : \Omega \to \Omega'$$

a random variable and

$$\mathbb{P}_{\mathbf{X}} : \mathcal{S}' \to [0, 1], \quad A' \mapsto \mathbb{P}(\{\omega \in \Omega; \, \mathbf{X}(\omega) \in A'\})$$

*the **image measure** of* $\mathbf{X}$*, also known as the **distribution** of* $\mathbf{X}$*, then:*

$$\mathbb{S}_{\mathbb{P}} \geq \mathbb{S}_{\mathbb{P}_{\mathbf{X}}}.$$

Let P' be a partition of Ω' from events, then:

$$-\sum_{E' \in P'} \mathbb{P}_{\mathbf{X}}(E') \, \text{ld}(\mathbb{P}_{\mathbf{X}}(E')) = - \sum_{E \in \{\mathbf{X}^{-1}(E'); \, E' \in P'\}} \mathbb{P}(E) \, \text{ld}(\mathbb{P}(E)).$$

Since $\{\mathbf{X}^{-1}(E'); \, E' \in P'\}$ is a partition of Ω from events, it follows

$$\mathbb{S}_{\mathbb{P}} \geq \mathbb{S}_{\mathbb{P}_{\mathbf{X}}}.$$

q.e.d

Theorem 7.5 is a generalization of theorem and definition 3.2.

In stochastic control theory, one generally has a probability space $(\Omega, \mathcal{S}, \mathbb{P})$ with $|\Omega| > |\mathbb{N}|$ given (this probability space represents the total information about the system to be controlled; often the possible results consist of functions (for example potential flight paths) and random variables

$$\mathbf{X}_t : \Omega \to \mathbb{R}^n, \quad n \in \mathbb{N}, \quad t \in [0, \infty),$$

representing the observation of the system at time t. Now let

$$S_t := \bigcap_{\mathcal{F} \subseteq S} \mathcal{F},$$

where each $\mathcal{F}$ represents a σ-field over Ω such that X_t-$\mathcal{F}$-$\mathcal{B}^n$-measurable, then S_t is the smallest σ-field over Ω, so that X_t-S_t-$\mathcal{B}^n$-measurable. Since the entropy $\mathbb{S}_{\mathbb{P}_{X_t}}$ for each $t \in [0, \infty)$ depends on the σ-field S_t (more precisely: on the partitions of Ω contained therein) and the probability measure common to all random variables $\mathbb{P}$, S_t is also used as a measure for the average amount of information from $(\mathbb{R}^n, \mathcal{B}^n, \mathbb{S}_{\mathbb{P}_{X_t}})$. This is particularly useful when

$$\mathbb{S}_{\mathbb{P}_{X_t}} = \infty \quad \text{for all} \quad t \in [0, \infty),$$

but still want to compare the probability spaces $(\mathbb{R}^n, \mathcal{B}^n, \mathbb{S}_{\mathbb{P}_{X_t}})$ for different $t \in [0, \infty)$ in terms of information theory. For example, one can interpret

$$S_{t_1} \subseteq S_{t_2}$$

in such a way that the observation of the system at time t_2 provides at least the information obtained by observing the system at time t_1.

7.3 Entropy in Dynamic Systems

The most important application of the Shannon entropy of partitions of Ω from events is given by the information-theoretical analysis of dynamic systems. Since the mathematical treatment of dynamic systems is a very extensive field, we will only present the essential ideas and refer to introductory literature on [EinSch14] and [Denker05]; specifically, the use of entropy to investigate dynamic systems is extensively treated in [Down11].

Let Ω be a non-empty set and

$$d : \Omega \times \Omega \to \mathbb{R}_0^+, \quad (x, y) \mapsto d(x, y)$$

an illustration, then d is referred to as a **metric** (on Ω) if the following conditions are met:

(M1) $d(x, y) = 0 \iff x = y$,
(M2) $d(x, y) = d(y, x)$ for all $x, y \in \Omega$,
(M3)   Triangle inequality:

$$d(x, z) \leq d(x, y) + d(y, z) \quad \text{for all} \quad x, y, z \in \Omega.$$

A **metric space** (Ω, d) is a pair consisting of a non-empty set Ω and a metric d on Ω. The value $d(x, y)$ is also referred to as the distance between x and y. Now let $M \subseteq \Omega$. If for every $x \in M$ there is an $\epsilon > 0$ with

$$\{y \in \Omega; \ d(x, y) < \epsilon\} \subseteq M,$$

then M is referred to as an **open set** (with respect to d). The sets Ω and $\emptyset$ are open sets. The σ-field $\mathcal{B}$ generated by all open sets over Ω is called **Borel σ-field**. If I is a non-empty index set and $M_i, i \in I$, are open subsets of Ω with

$$\Omega \subseteq \bigcup_{i \in I} M_i,$$

then $\{M_i;\ i \in I\}$ is a **open cover** of Ω is called. The set Ω is called **compact** if for every open cover of Ω, only finitely many sets from this open cover are sufficient to cover Ω. If Ω is compact, then (Ω, d) **compact metric space** is called. Now let (Ω, d) be a compact metric space and let furthermore

$$T : \Omega \to \Omega$$

be a $\mathcal{B}$-$\mathcal{B}$-measurable mapping, then the triple $(\Omega, \mathcal{B}, T)$ is referred to as a **dynamic system**. If one chooses an $x_1 \in \Omega$, then the set

$$\{T^k(x_1);\ k \in \mathbb{N}_0\} = \{x_1, T(x_1), T^2(x_1), \ldots\}$$

orbit or **trajectory** of the dynamic system is called. This implies the agreement

$$T^k = T \circ T^{k-1},\ k \in \mathbb{N},\ \text{and}\ T^0 : \Omega \to \Omega,\ x \mapsto x.$$

If, for example, one examines the dynamic system

$$([0, 1], \mathcal{B}, T)$$

with

$$T : [0, 1] \to [0, 1], \quad x \mapsto 4x(1 - x) \quad \text{(logistic transformation)},$$

where the metric, as usual with real numbers, is given by

$$d : [0, 1] \times [0, 1] \to \mathbb{R}, \quad (x, y) \mapsto |x - y|,$$

one can initially consider various orbits. For $x_1 = 1$ one obviously obtains the orbit

$$\{1, 0, 0, \ldots\}.$$

The situation is completely different for $x_1 = 0.9999$, as Fig. 7.1 shows.

It can also happen that an orbit transitions into an alternating state (Fig. 7.2); furthermore, there are also orbits that eventually become constant (Fig. 7.3).

In addition to the analysis of individual orbits, a probabilistic analysis of a dynamic system is also possible. For this, we supplement the underlying measurable space $(\Omega, \mathcal{B})$ with a probability measure $\mathbb{P}$ defined on $\mathcal{B}$ and now examine the sequence of image measures

$$\mathbb{P}, \mathbb{P}_T, \mathbb{P}_{T^2}, \ldots$$

Is there a probability measure $\mathbb{P}_*$ such that the above sequence is constant (i.e., $\mathbb{P}_{T^k} = \mathbb{P}_*$ for all $k \in \mathbb{N}_0$), this probability measure is referred to as **invariant measure**.

Another classic approach to the analysis of dynamic systems directly refers to the Shannon entropy of partitions of Ω from events according to definition 7.2. If $P_{\mathcal{B}}$ is a partition of Ω from events, then for each $k \in \mathbb{N}_0$ the set

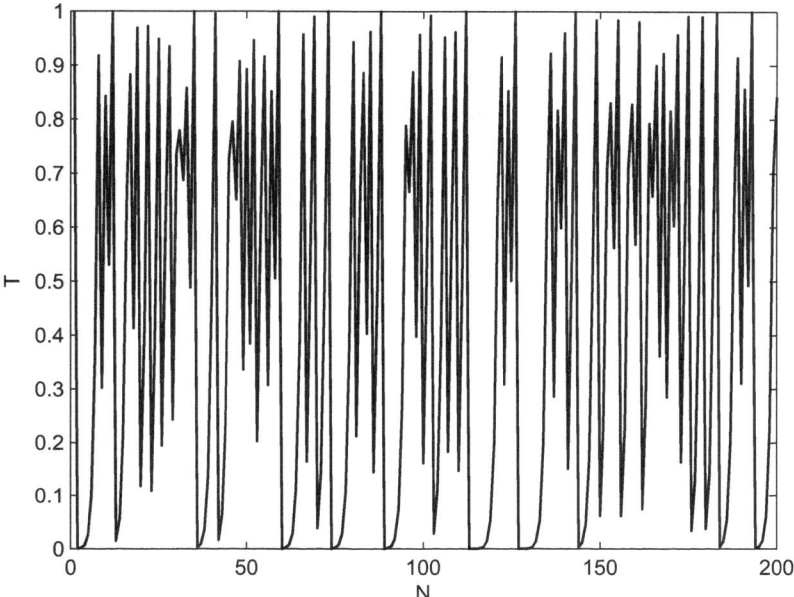

Fig. 7.1   Logistic transformation, orbit for $x_1 = 0.9999$

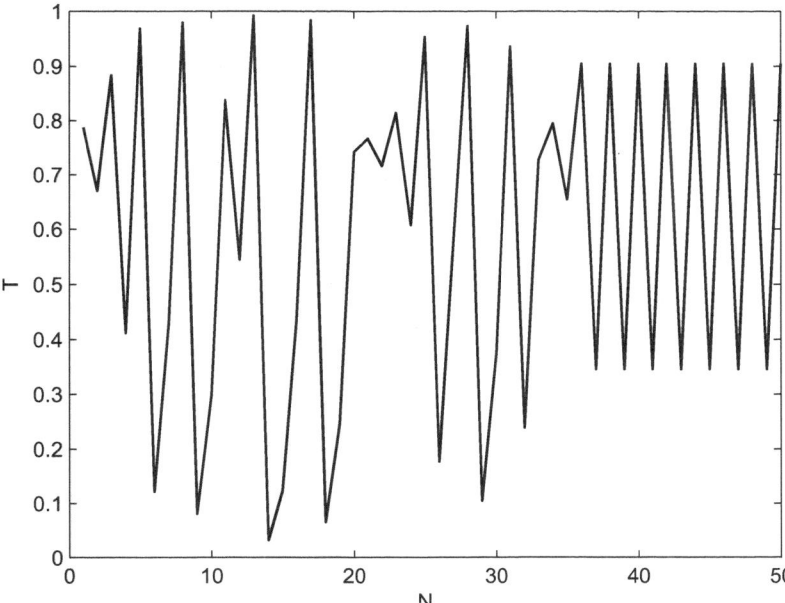

Fig. 7.2   Logistic transformation, finally alternating orbit

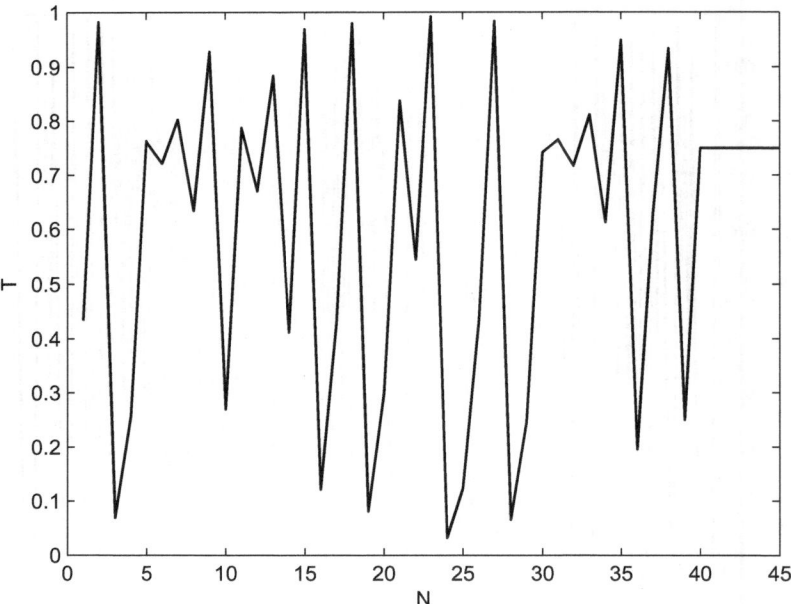

Fig. 7.3 Logistic transformation, finally constant orbit

$$T^{-k}(P_{\mathcal{B}}) := \{T^{-k}(A);\ A \in P_{\mathcal{B}}\} \setminus \{\emptyset\}$$

is also a partition of Ω from events (with $T^{-k}(A) = A$ for $k=0$, $A \in \mathcal{B}$). Furthermore, for each $N \in \mathbb{N}$ the set

$$\bigvee_{k=0}^{N} T^{-k}(P_{\mathcal{B}}) := \left\{\bigcap_{k=0}^{N} T^{-k}(A_k);\ A_k \in P_{\mathcal{B}}\right\} \setminus \{\emptyset\}$$

is a partition of Ω from events. If you now choose sets

$$A_0, A_1, \ldots, A_N \in P_{\mathcal{B}},$$

there is exactly one set

$$M \in \bigvee_{k=0}^{N} T^{-k}(P_{\mathcal{B}})$$

with

$$(x \in A_0) \wedge (T(x) \in A_1) \wedge \ldots \wedge \left(T^N(x) \in A_N\right) \quad \Longleftrightarrow \quad x \in M$$

and conversely, for each set

$$M_1 \in \bigvee_{k=0}^{N} T^{-k}(P_{\mathcal{B}})$$

there is exactly one choice of sets $B_0, B_1, \ldots, B_N \in P_{\mathcal{B}}$ with

$$x \in M_1 \quad \Longleftrightarrow \quad (x \in B_0) \wedge (T(x) \in B_1) \wedge \ldots \wedge \left(T^N(x) \in B_N\right).$$

For example, if we choose the logistic transformation

$$P_{\mathcal{B}} = \left\{ \left[0, \frac{1}{2}\right), \left[\frac{1}{2}, 1\right] \right\},$$

we get for $N=1$:

$$\bigvee_{k=0}^{1} T^{-k}(P_{\mathcal{B}}) = \left\{ \left[0, \frac{1}{2} - \frac{1}{2}\sqrt{1 - \frac{1}{2}}\right), \left[\frac{1}{2} - \frac{1}{2}\sqrt{1 - \frac{1}{2}}, \frac{1}{2}\right) \right.$$
$$\left. \left[\frac{1}{2}, \frac{1}{2} + \frac{1}{2}\sqrt{1 - \frac{1}{2}}\right], \left(\frac{1}{2} + \frac{1}{2}\sqrt{1 - \frac{1}{2}}, 1\right] \right\}.$$

The choice

$$M_1 = \left(\frac{1}{2} + \frac{1}{2}\sqrt{1 - \frac{1}{2}}, 1\right]$$

yields, for example:

$$x \in M_1 \quad \Longleftrightarrow \quad \left(x \in \left[\frac{1}{2}, 1\right]\right) \wedge \left(T(x) \in \left[0, \frac{1}{2}\right)\right).$$

Conversely, the choice

$$A_0 = A_1 = \left[0, \frac{1}{2}\right)$$

yields the set M with

$$\left(x \in \left[0, \frac{1}{2}\right)\right) \wedge \left(T(x) \in \left[0, \frac{1}{2}\right)\right) \quad \Longleftrightarrow \quad x \in M = \left[0, \frac{1}{2} - \frac{1}{2}\sqrt{1 - \frac{1}{2}}\right).$$

If one has a given dynamic system $(\Omega, \mathcal{B}, T)$ and a probability measure $\mathbb{P}$ defined on $\mathcal{B}$, one can calculate the entropy per orbit length

$$\mathbb{S}_{\mathbb{P}, P_{\mathcal{B}}, N} := -\frac{\displaystyle\sum_{A \in \bigvee_{k=0}^{N-1} T^{-k}(P_{\mathcal{B}})} \mathbb{P}(A)\, \mathrm{ld}(\mathbb{P}(A))}{N}$$

for $N \in \mathbb{N}$ and $P_{\mathcal{B}}$.

In general, for a dynamic system $(\Omega, \mathcal{B}, T)$ and a partition $P_{\mathcal{B}}$ of Ω from events, one can show that the limit

$$\mathbb{S}_{\mathbb{P}, P_{\mathcal{B}}} := \lim_{N \to \infty} \mathbb{S}_{\mathbb{P}, P_{\mathcal{B}}, N}$$

exists if one assumes an invariant measure $\mathbb{P}_*$.

The size $\mathbb{S}_{\mathbb{P}, P_{\mathcal{B}}}$ is referred to as the **entropy of the transformation T with respect to $P_{\mathcal{B}}$**. For the logistic transformation with a probability measure $\mathbb{P}$ clearly defined by

$$\mathbb{P}((a, b]) = b - a \quad \text{für} \quad a, b \in [0, 1], \quad a < b$$

and for

$$P_{\mathcal{B}} = \left\{ \left[0, \frac{1}{2}\right), \left[\frac{1}{2}, 1\right] \right\},$$

the entropies per orbit length are shown as in Fig. 7.4.

In order to classify the transformation T independently of the chosen partition of Ω from events, one considers with $\Pi_{\mathcal{B}}$ the set of all partitions of Ω from events and one refers to the size

$$\mathbb{S}_{\mathbb{P}_*}(T) := \sup_{P_{\mathcal{B}} \in \Pi_{\mathcal{B}}} \mathbb{S}_{\mathbb{P}_*, P_{\mathcal{B}}}$$

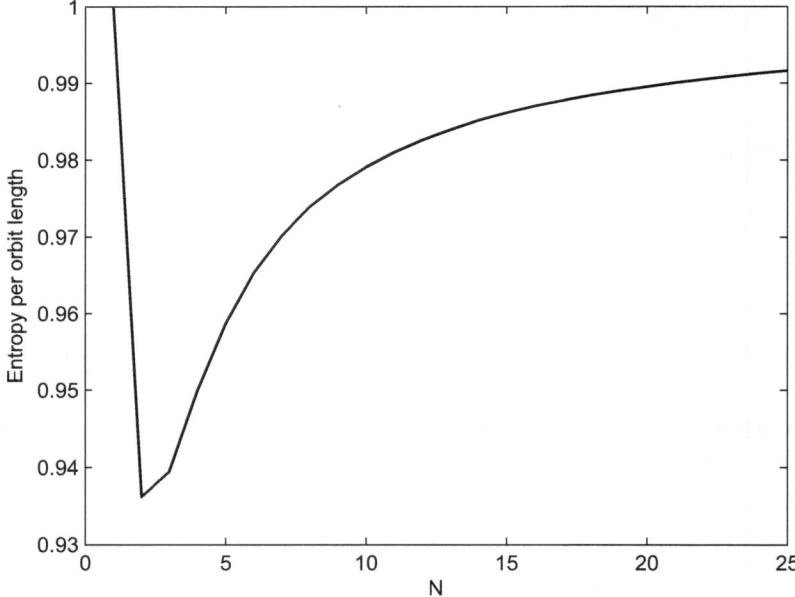

Fig. 7.4 $\mathbb{S}_{\mathbb{P}, \{[0, \frac{1}{2}), [\frac{1}{2}, 1]\}, N}$, logistic transformation

as **Kolmogorov-Sinai entropy** of T, where $\mathbb{P}_*$ represents an invariant measure.

Since an invariant measure $\mathbb{P}_*$ is used in the calculation of the Kolmogorov-Sinai entropy, we now investigate the question under which conditions invariant measures exist. We will forego the essentially functional-analytical proof and refer again to [EinSch14] and [Denker05].

Let (Ω, d) be a compact metric space, $\mathcal{O}$ the set of all d-open subsets of Ω, and $\mathcal{B}$ the Borel σ-algebra generated by $\mathcal{O}$. Furthermore, let

$$T : \Omega \to \Omega$$

be a $\mathcal{B}$-$\mathcal{B}$-measurable mapping and thus a $(\Omega, \mathcal{B}, T)$ dynamic system. Since it is obvious that for a $\hat{x} \in \Omega$

$$\mathbb{P} : \mathcal{B} \to [0, 1], \quad A \mapsto \begin{cases} 1 & \text{if } \hat{x} \in A \\ 0 & \text{if } \hat{x} \notin A \end{cases}$$

represents a probability measure, the set $\mathfrak{W}$ of all probability measures on $\mathcal{B}$ is not empty.

As the next step, we define a metric on $\mathfrak{W}$. For this purpose, we examine the set $C(\Omega, \mathbb{R})$ of all continuous functions

$$f : \Omega \to \mathbb{R},$$

that is, all functions f such that the pre-image

$$f^{-1}(O) := \{x \in \Omega; f(x) \in O\}$$

of an open subset O of $\mathbb{R}$ is open (i.e., an element of $\mathcal{O}$). Through the mapping

$$\| \cdot \|_\infty : C(\Omega, \mathbb{R}) \to \mathbb{R}_0^+, \quad f \mapsto \max_{x \in \Omega}\{|f(x)|\}$$

a norm on $C(\Omega, \mathbb{R})$ is defined. An important result of functional analysis now states that the normed space $(C(\Omega, \mathbb{R}), \| \cdot \|_\infty)$ represents a separable Banach space, that is

(1) every Cauchy sequence consisting of elements from $C(\Omega, \mathbb{R})$ with respect to the given norm $\| \cdot \|_\infty$ converges to an element from $C(\Omega, \mathbb{R})$ (Banach space) and that

(2) there is a countable subset $D \subset C(\Omega, \mathbb{R})$ with

$$\bigcap_{\substack{D \subseteq M^c \\ M \in \mathcal{O}}} M^c = C(\Omega, \mathbb{R}) \quad \text{(separability)}.$$

The set D described in (2) is referred to as a **countable dense subset of** $C(\Omega, \mathbb{R})$. Since D is countable, we can count the elements of D, i.e., write them in the form

$$D = \{d_1, d_2, \ldots\}$$

and furthermore, we can assume that the zero function

$$f_0 : \Omega \rightarrow \mathbb{R}, \quad x \mapsto 0$$

does not belong to D.

A metric $d_{\mathfrak{W}}$ on $\mathfrak{W}$ is now obtained by

$$d_{\mathfrak{W}} : \mathfrak{W} \times \mathfrak{W} \rightarrow \mathbb{R}_0^+, \quad (\mathbb{P}_1, \mathbb{P}_2) \mapsto \sum_{i=1}^{\infty} \frac{\left| \int d_i d\mathbb{P}_1 - \int d_i d\mathbb{P}_2 \right|}{2^i \|d_i\|_{\infty}},$$

where $\int d_i d\mathbb{P}_1$ and $\int d_i d\mathbb{P}_2$ represent integrals that are introduced in Sect. 9.1.

This metric defines the convergence of probability measures from $\mathfrak{W}$ by

$$\lim_{n \to \infty} \mathbb{P}_n = \mathbb{P} \quad :\Longleftrightarrow \quad \lim_{n \to \infty} d_{\mathfrak{W}}(\mathbb{P}_n, \mathbb{P}) = 0.$$

If now $\{\mathbb{P}_k\}_{k \in \mathbb{N}}$ is a sequence of probability measures from $\mathfrak{W}$, we generate another sequence of probability measures from $\mathfrak{W}$ by

$$\hat{\mathbb{P}}_k : \mathcal{B} \rightarrow [0,1], \quad A \mapsto \frac{1}{k} \sum_{j=0}^{k-1} \mathbb{P}_{k, T^j}(A), \quad k \in \mathbb{N},$$

where $\mathbb{P}_{k, T^j}$ represents the image measure of T^j with respect to $\mathbb{P}_k$. The special thing about this approach is that $\{\hat{\mathbb{P}}_k\}_{k \in \mathbb{N}}$ converges

$$\lim_{k \to \infty} \hat{\mathbb{P}}_k = \mathbb{P}_*$$

and that an invariant measure is given by the limit $\mathbb{P}_*$.

Example 7.6 Let's return to the logistic transformation $([0,1], \mathcal{B}, T)$ with

$$T : [0,1] \rightarrow [0,1], \quad x \mapsto 4x(1-x).$$

If one chooses the constant sequence

$$\mathbb{P}_k : \mathcal{B} \rightarrow [0,1], \quad A \mapsto \begin{cases} 1 & \text{if } \frac{3}{4} \in A \\ 0 & \text{if } \frac{3}{4} \notin A \end{cases}, \quad k \in \mathbb{N},$$

then

$$\hat{\mathbb{P}}_k = \mathbb{P}_k \quad \text{for all} \quad k \in \mathbb{N}$$

applies and

$$\mathbb{P}_* : \mathcal{B} \rightarrow [0,1], \quad A \mapsto \begin{cases} 1 & \text{if } \frac{3}{4} \in A \\ 0 & \text{if } \frac{3}{4} \notin A \end{cases}$$

is an invariant measure. The invariant measure

$$\mathbb{P}_* : \mathcal{B} \to [0,1], \quad A \mapsto \begin{cases} \frac{1}{2} & \text{if } \frac{5+\sqrt{5}}{8} \in A, \ \frac{5-\sqrt{5}}{8} \notin A \\ \frac{1}{2} & \text{if } \frac{5-\sqrt{5}}{8} \in A, \ \frac{5+\sqrt{5}}{8} \notin A \\ 0 & \text{if } \frac{5+\sqrt{5}}{8} \notin A, \ \frac{5-\sqrt{5}}{8} \notin A \\ 1 & \text{if } \frac{5+\sqrt{5}}{8} \in A, \ \frac{5-\sqrt{5}}{8} \in A \end{cases}$$

takes into account the fact that there is an alternating orbit (see Fig. 7.2).   ◁

The following theorem summarizes the knowledge about invariant measures.

▶   **Theorem 7.7 (Kryloff-Bogoliuboff)** *Let* $(\Omega, \mathcal{B}, T)$ *be a dynamic system,* $\mathfrak{W}$ *the set of all probability measures on* $\mathcal{B}$ *and* $\mathfrak{W}_T \subseteq \mathfrak{W}$ *the set of all invariant measures, then* $\mathfrak{W}_T$ *is a non-empty, compact and convex subset of* $\mathfrak{W}$.

The convexity of the set $\mathfrak{W}_T$ means that with two invariant measures $\mathbb{P}_*^1$ and $\mathbb{P}_*^2$ every probability measure

$$\alpha \mathbb{P}_*^1 + (1 - \alpha)\mathbb{P}_*^2 : \mathcal{B} \to [0,1], \quad A \mapsto \alpha \mathbb{P}_*^1(A) + (1 - \alpha)\mathbb{P}_*^2(A) \quad 0 \le \alpha \le 1$$

is an invariant measure. Now, if for an invariant measure $\mathbb{P}_*$:

$$\mathbb{P}_* = \alpha \mathbb{P}_*^1 + (1 - \alpha)\mathbb{P}_*^2, \quad 0 < \alpha < 1 \quad \text{with} \quad \mathbb{P}_*^1(A), \mathbb{P}_*^2(A) \in \mathfrak{W}_T$$

exactly when

$$\mathbb{P}_* = \mathbb{P}_*^1 = \mathbb{P}_*^2,$$

then it is called $\mathbb{P}_*$ an **extremal point** of $\mathfrak{W}_T$. Since $\mathfrak{W}_T$ is compact (and therefore closed), $\mathfrak{W}_T$ has at least one extremal point. These extremal points are **ergodic probability measures**: If one chooses for a dynamic system $(\Omega, \mathcal{B}, T)$ and an invariant probability measure $\mathbb{P}_*$ a set $M \in \mathcal{B}$ with $T^{-1}(M) = M$ (a so-called T-invariant set, e.g. Ω and $\emptyset$; orbits with starting point from M remain in M), this probability measure is called ergodic, if:

$$\mathbb{P}_*(M) = 0 \quad \text{oder} \quad \mathbb{P}_*(M) = 1.$$

This means that there is no partition $\{P_1, P_2\}$ of Ω from T-invariant events with

$$0 < \mathbb{P}_*(P_1) < 1, \quad 0 < \mathbb{P}_*(P_2) < 1.$$

So if one wants to decompose the dynamic system $(\Omega, \mathcal{B}, T)$ through T-invariant sets into non-trivial subsystems to analyze them, this is not possible on the basis of an ergodic probability measure, since on the one hand the full complexity is preserved ($\mathbb{P}_*(P_1) = 1$) and on the other hand one considers $\mathbb{P}_*$-null sets ($\mathbb{P}_*(P_2) = 0$).

For the logistic transformation, the invariant measure

$$\mathbb{P}_* : \mathcal{B} \to [0,1], \quad A \mapsto \begin{cases} 1 & \text{if } \frac{3}{4} \in A \\ 0 & \text{if } \frac{3}{4} \notin A \end{cases}$$

is ergodic, because for a T-invariant set M with $\frac{3}{4} \in M$ applies $\mathbb{P}_*(M) = 1$, otherwise $\mathbb{P}_*(M) = 0$. Also

$$\mathbb{P}_* : \mathcal{B} \to [0,1], \quad A \mapsto \begin{cases} \frac{1}{2} & \text{if } \frac{5+\sqrt{5}}{8} \in A, \ \frac{5-\sqrt{5}}{8} \notin A \\ \frac{1}{2} & \text{if } \frac{5-\sqrt{5}}{8} \in A, \ \frac{5+\sqrt{5}}{8} \notin A \\ 0 & \text{if } \frac{5+\sqrt{5}}{8} \notin A, \ \frac{5-\sqrt{5}}{8} \notin A \\ 1 & \text{if } \frac{5+\sqrt{5}}{8} \in A, \ \frac{5-\sqrt{5}}{8} \in A \end{cases}$$

is ergodic, because every T-invariant set M with $\frac{5+\sqrt{5}}{8} \in M$ also contains $\frac{5-\sqrt{5}}{8}$ and vice versa.

The information-theoretical analysis of dynamic systems must remain incomplete for now, as we do not yet have the concept of the density function and the associated differential entropy available. We will therefore return to this topic in Sect. 9.4.

Stationary Information Sources

8

8.1   Cylinder Sets and Projections

Starting from a non-empty set A, which we refer to as a character set and which may contain at least two and at most finitely many elements (so-called characters), in communication technology one often starts from a transmitter who sends a character $a \in A$ at every point in time $t \in \mathbb{Z}$. At any fixed point in time $t_0 \in \mathbb{Z}$, the transmitter has thus already sent infinitely many characters and will also send infinitely many characters after t_0. This unrealistic assumption serves to capture all practically relevant cases in a single mathematical model and not to have to design a separate mathematical model for each special case; we will come back to this.

We denote the set Ω of all messages as the set of all possible character sequences that can be generated by the transmitter:

$$\Omega = A^{\mathbb{Z}} := \{f : \mathbb{Z} \to A\}.$$

This set now acts as the outcome set of a probability space. Since $|A^{\mathbb{Z}}| > |\mathbb{N}|$, caution is required when choosing the events (given by a σ-field S); the thoughtless choice of the power set of Ω can lead to a strong impairment in the choice of probability measures. In practice, the following procedure has proven useful: For each $k \in \mathbb{N}, \{i_1, \ldots, i_k\} \subset \mathbb{Z}$ and $\mathbf{a} \in A^k$ one considers the sets

$$\mathcal{Z}_{k,\{i_1,\ldots,i_k\},\mathbf{a}} := \{f \in A^{\mathbb{Z}}; f(i_j) = a_j, j = 1, \ldots, k\}$$

and the set

$$\mathcal{Z} := \{\mathcal{Z}_{k,\{i_1,\ldots,i_k\},\mathbf{a}}; \ k \in \mathbb{N}, \{i_1, \ldots, i_k\} \subset \mathbb{Z}, \mathbf{a} \in A^k\} \cup \{\emptyset\} \cup \{\Omega\}$$

(at finitely many fixed points in time, fixed characters were sent). As σ-field S over $A^{\mathbb{Z}}$, the smallest σ-field is now chosen that contains all sets from $\mathcal{Z}$ (called cylinder sets):

S. Schäffler, *Mathematics of Information*, Mathematics Study Resources 9, https://doi.org/10.1007/978-3-662-69102-1_8

$$S = \bigotimes_{i \in \mathbb{Z}} \mathcal{P}(A) := \sigma(\mathcal{Z}).$$

The set $\mathcal{Z}$ of all cylinder sets has the structure of a **Semi-Ring** $\mathcal{SR}$ over Ω:

(SR1) $\emptyset \in \mathcal{SR}$.

(SR2) From $A, B \in \mathcal{SR}$ follows $A \cap B \in \mathcal{SR}$.

(SR3) For any two sets $A, B \in \mathcal{SR}$ is $A \setminus B$ a finite union of pairwise disjoint sets from $\mathcal{SR}$.

Now, if we have a mapping w on $\mathcal{Z}$ with:

(i) $w : \mathcal{Z} \to [0, 1]$,

(ii) $w(\emptyset) = 0$,

(iii) for any countably many $A, A_1, A_2, \ldots \in \mathcal{Z}$ with $A \subseteq \bigcup_{i=1}^{\infty} A_i$ applies:

$$w(A) \leq \sum_{i=1}^{\infty} w(A_i),$$

given, then with the extension theorem 1.53 from [Klenke05], we obtain a unique probability measure $\mathbb{P}$ on $\bigotimes_{i \in \mathbb{Z}} \mathcal{P}(A)$ with

$$\mathbb{P}(A) = w(A) \quad \text{for all} \quad A \in \mathcal{Z}.$$

Now we consider the projections

$$p_i : A^{\mathbb{Z}} \to A, \quad f \mapsto f(i), \quad i \in \mathbb{Z}.$$

Since for each $M \subseteq A$ applies:

$$p_i^{-1}(M) = \bigcup_{m \in M} \mathcal{Z}_{1, \{i\}, m} \quad (\text{where} \quad p_i^{-1}(\emptyset) := \emptyset)$$

and since

$$\mathcal{Z}_{k, \{i_1, \ldots, i_k\}, \mathbf{a}} = \bigcap_{j=1}^{k} p_{i_j}^{-1}(\{a_j\}),$$

is $\bigotimes_{i \in \mathbb{Z}} \mathcal{P}(A)$ is also the smallest σ-field over $A^{\mathbb{Z}}$, such that every projection p_i, $i \in \mathbb{Z}$, $\bigotimes_{i \in \mathbb{Z}} \mathcal{P}(A)$-$\mathcal{P}(A)$-measurable is. This results in another great advantage of the σ-field $\bigotimes_{i \in \mathbb{Z}} \mathcal{P}(A)$.

▶ Theorem 8.1 (Closure under bijections) *Let A be a set of characters, $\Omega = A^{\mathbb{Z}}$,*
$\mathcal{S} = \bigotimes\limits_{i \in \mathbb{Z}} \mathcal{P}(A)$, $B : \mathbb{Z} \to \mathbb{Z}$ *a bijection and*

$$\tau_B : \bigotimes\limits_{i \in \mathbb{Z}} \mathcal{P}(A) \to \mathcal{P}\left(A^{\mathbb{Z}}\right), \quad S \mapsto \{g : \mathbb{Z} \to A, \ k \mapsto f(B(k)); f \in S\},$$

then for every set $S \in \bigotimes\limits_{i \in \mathbb{Z}} \mathcal{P}(A)$:

◁
$$\tau_B(S) \in \bigotimes\limits_{i \in \mathbb{Z}} \mathcal{P}(A).$$

We examine the set system

$$\mathcal{F} = \left\{ S \in \bigotimes\limits_{i \in \mathbb{Z}} \mathcal{P}(A); \ \tau_B(S) \in \bigotimes\limits_{i \in \mathbb{Z}} \mathcal{P}(A) \right\}.$$

Since $\mathcal{F} \subseteq \bigotimes\limits_{i \in \mathbb{Z}} \mathcal{P}(A)$, it remains to show that $\bigotimes\limits_{i \in \mathbb{Z}} \mathcal{P}(A) \subseteq \mathcal{F}$.
Let

$$C \in \{p_j^{-1}(N); \ j \in \mathbb{Z}, N \in \mathcal{P}(A)\},$$

be, then there exists an $i \in \mathbb{Z}$ and an $M \in \mathcal{P}(A)$ with

$$C = p_i^{-1}(M) = \{f \in A^{\mathbb{Z}}; \ f(i) \in M\}$$

and

$$\begin{aligned}
\tau_B(C) &= \{g : \mathbb{Z} \to A, \ k \mapsto f(B(k)); f \in C\} \\
&= \{g : \mathbb{Z} \to A; \ g(B^{-1}(i)) \in M\} \\
&= p_{B^{-1}(i)}^{-1}(M).
\end{aligned}$$

Therefore,

$$\{p_j^{-1}(N); \ j \in \mathbb{Z}, N \in \mathcal{P}(A)\} \subseteq \mathcal{F}.$$

applies. Due to

$$\bigotimes\limits_{i \in \mathbb{Z}} \mathcal{P}(A) = \sigma\left(\{p_j^{-1}(N); \ j \in \mathbb{Z}, N \in \mathcal{P}(A)\}\right),$$

it remains to show that $\mathcal{F}$ is a σ-field.

(i) Since $\tau_B(\Omega) = \{g : \mathbb{Z} \to A, \ k \mapsto f(B(k)); f \in A^{\mathbb{Z}}\} = \Omega \in \bigotimes\limits_{i \in \mathbb{Z}} \mathcal{P}(A), \Omega \in \mathcal{F}$ is.

(ii) Since with $S \in \mathcal{F}$ applies:

$$\tau_B(S^c) = \{g : \mathbb{Z} \to A, \, k \mapsto f(B(k)); f \in S^c\}$$
$$= \{g : \mathbb{Z} \to A, \, k \mapsto f(B(k)); f \in S\}^c$$
$$= \tau_B(S)^c \in \bigotimes_{i \in \mathbb{Z}} \mathcal{P}(A),$$

is $S^c \in \mathcal{F}$.

(iii) Since with $S_i \in \mathcal{F}, i \in \mathbb{N}$ applies:

$$\tau_B\left(\bigcup_{i=1}^{\infty} S_i\right) = \left\{g : \mathbb{Z} \to A, \, k \mapsto f(B(k)); f \in \bigcup_{i=1}^{\infty} S_i\right\}$$
$$= \bigcup_{i=1}^{\infty} \{g : \mathbb{Z} \to A, \, k \mapsto f(B(k)); f \in S_i\}$$
$$= \bigcup_{i=1}^{\infty} \tau_B(S_i) \in \bigotimes_{i \in \mathbb{Z}} \mathcal{P}(A),$$

is $\bigcup_{i=1}^{\infty} S_i \in \mathcal{F}$.

Now we are able to define (stationary) information sources.

Definition 8.2 ((stationary) Information Source) Let A be a character set (so: $2 \le |A| < \infty$), a probability space

$$\left(A^{\mathbb{Z}}, \bigotimes_{i \in \mathbb{Z}} \mathcal{P}(A), \mathbb{P}\right)$$

is referred to as an **information source**. An information source is called **stationary** if for every bijection

$$B_q : \mathbb{Z} \to \mathbb{Z}, \quad i \mapsto i + q, \quad q \in \mathbb{Z}$$

the following applies:

◁

$$\mathbb{P}(S) = \mathbb{P}(\tau_{B_q}(S)) \quad \text{for all} \quad S \in \bigotimes_{i \in \mathbb{Z}} \mathcal{P}(A).$$

The probabilistic statements about a stationary information source are thus time-invariant.

Since for every $r \in \mathbb{N}$ the following applies:

$$\tau_{B_r}(S) = \underbrace{(\tau_{B_1} \circ \ldots \circ \tau_{B_1})}_{r\text{-mal}}(S) \quad \text{for all} \quad S \in \bigotimes_{i \in \mathbb{Z}} \mathcal{P}(A)$$

and

$$S = \underbrace{(\tau_{B_1} \circ \ldots \circ \tau_{B_1})}_{\text{r-mal}}(\tau_{B_{-r}}(S)) \quad \text{for all} \quad S \in \bigotimes_{i \in \mathbb{Z}} \mathcal{P}(A),$$

an information source is already stationary if with

$$B_1 : \mathbb{Z} \to \mathbb{Z}, \quad i \mapsto i + 1$$

the following applies:

$$\mathbb{P}(S) = \mathbb{P}(\tau_{B_1}(S)) \quad \text{for all} \quad S \in \bigotimes_{i \in \mathbb{Z}} \mathcal{P}(A).$$

If you now want to analyze a finite message consisting of K characters, where the characters are sent at the times $t_1 < \ldots < t_K \in \mathbb{Z}$, this can be done based on the information source $\left(A^{\mathbb{Z}}, \bigotimes_{i \in \mathbb{Z}} \mathcal{P}(A), \mathbb{P} \right)$ by analyzing the mapping

$$\mathbf{X} : A^{\mathbb{Z}} \to A^K, \quad f \mapsto (f(t_1), \ldots, f(t_K)),$$

because using the power set $\mathcal{P}(A^K)$ of A^K is $\mathbf{X} \bigotimes_{i \in \mathbb{Z}} \mathcal{P}(A)$-$\mathcal{P}(A^K)$-measurable. It is therefore not necessary to define a new probability space for each new finite message, but all possible finite messages can be analyzed and compared with each other by suitable random variables based on **one** probability space

$$\left(A^{\mathbb{Z}}, \bigotimes_{i \in \mathbb{Z}} \mathcal{P}(A), \mathbb{P} \right)$$

.

Example 8.3 Let A be a character set and

$$\mathbb{P}_A : \mathcal{P}(A) \to [0, 1]$$

a probability measure on $\mathcal{P}(A)$. Further, with $J \subset \mathbb{Z}$, $1 \leq |J| < \infty$ and $(a_1, \ldots, a_{|J|}) \in A^{|J|}$

$$P : \mathcal{Z} \to [0, 1], \quad \{f \in A^{\mathbb{Z}}, f(i_j) = a_j; \ \{i_1, \ldots, i_{|J|}\} = J\} \mapsto \prod_{j=1}^{|J|} \mathbb{P}_A(\{a_j\}),$$

$$\Omega \mapsto 1,$$

$$\emptyset \mapsto 0,$$

there exists exactly one probability measure

$$\mathbb{P} : \bigotimes_{i \in \mathbb{Z}} \mathcal{P}(A) \to [0, 1]$$

with $\mathbb{P}_{|\mathcal{Z}} = P$.

The information source $\left(A^{\mathbb{Z}}, \bigotimes_{i \in \mathbb{Z}} \mathcal{P}(A), \mathbb{P} \right)$ is stationary.   ◁

8.2   Entropy per Character

The decisive characteristic for a stationary information source with character set A is the entropy per character, which we will now introduce.

Theorem and Definition 8.4 (Entropy per Character) *Let A be a character set and*

$$\left(A^{\mathbb{Z}}, \bigotimes_{i \in \mathbb{Z}} \mathcal{P}(A), \mathbb{P} \right)$$

a stationary information source, then for every $N \in \mathbb{N}, i \in \mathbb{Z}$:

The mapping

$$\mathbf{X}_{N,i} : A^{\mathbb{Z}} \to A^N, \quad f \mapsto (f(i), f(i+1), \ldots, f(i+N-1))$$

is $\bigotimes_{i \in \mathbb{Z}} \mathcal{P}(A)$-$\mathcal{P}(A^N)$-measurable, where $\mathcal{P}(A^N)$ represents the power set of A^N.

For the image measures, the following applies:

$$\mathbb{P}_{\mathbf{X}_{N,j}} = \mathbb{P}_{\mathbf{X}_{N,k}} \quad \text{for all} \quad j, k \in \mathbb{Z}.$$

Furthermore, the limit

$$\bar{\mathbb{S}}_{\mathbb{P}} := \lim_{N \to \infty} \frac{\mathbb{S}_{\mathbb{P}_{\mathbf{X}_{N,i}}}}{N} \quad \text{independent of} \quad i \in \mathbb{Z}$$

*exists and is referred to as the **entropy per character** of a stationary information source with the unit $\left[\frac{\text{bit}}{\text{Zeichen}} \right]$.*   ◁

Example (see [HeHo74] for this) If $M \in \mathcal{P}(A^N)$ is the case, then

$$\mathbf{X}_{N,i}^{-1}(M) = \bigcup_{\mathbf{m} \in M} \mathbf{X}_{N,i}^{-1}(\{\mathbf{m}\}) = \bigcup_{\mathbf{m} \in M} \mathcal{Z}_{N,\{i,i+1,\ldots,i+N-1\},\mathbf{m}} \in \bigotimes_{i \in \mathbb{Z}} \mathcal{P}(A)$$

and thus the mapping $\mathbf{X}_{N,i} \bigotimes \mathcal{P}(A)$-$\mathcal{P}(A^N)$-measurable. For the image dimensions, we obtain with the bijection $F^{\mathbb{Z}}$

$$B_{k-j} : \mathbb{Z} \to \mathbb{Z}, \quad i \mapsto i + k - j :$$

$$\mathbb{P}_{\mathbf{X}_{N,j}}(M) = \mathbb{P}(\mathbf{X}_{N,j}^{-1}(M)) = \sum_{\mathbf{m}\in M} \mathbb{P}(\mathbf{X}_{N,j}^{-1}(\{\mathbf{m}\}))$$

$$= \sum_{\mathbf{m}\in M} \mathbb{P}(\{f \in A^{\mathbb{Z}}; \ f(j+i-1) = m_i, \ i = 1, \dots, N\})$$

$$= \sum_{\mathbf{m}\in M} \mathbb{P}(\{f \in A^{\mathbb{Z}}; \ f(B_{k-j}(j+i-1)) = m_i, \ i = 1, \dots, N\})$$

$$= \sum_{\mathbf{m}\in M} \mathbb{P}(\{f \in A^{\mathbb{Z}}; \ f(k+i-1) = m_i, \ i = 1, \dots, N\})$$

$$= \mathbb{P}_{\mathbf{X}_{N,k}}(M).$$

Now we consider for $N, M \in \mathbb{N}$ and $i \in \mathbb{Z}$ the probability space

$$(A^{N+M}, \mathcal{P}(A^{N+M}), \mathbb{P}_{\mathbf{X}_{N+M,i}}), \quad \text{the measurable space} \quad (A^N, \mathcal{P}(A^N))$$

and the random variable

$$\mathbf{X} : A^{N+M} \to A^N, \quad (m_1, \dots, m_{N+M}) \mapsto (m_1, \dots, m_N)$$

with the image measure $\mathbb{P}_{\mathbf{X}_{N+M,i},\mathbf{X}} : \mathcal{P}(A^N) \to [0,1]$.
For each $B \subseteq A^N$ it holds:

$$\mathbb{P}_{\mathbf{X}_{N+M,i},\mathbf{X}}(B) = \mathbb{P}_{\mathbf{X}_{N+M,i}}(\mathbf{X}^{-1}(B)) = \sum_{\mathbf{b}\in B} \mathbb{P}_{\mathbf{X}_{N+M,i}}(\mathbf{X}^{-1}(\{\mathbf{b}\}))$$

$$= \sum_{\mathbf{b}\in B} \mathbb{P}_{\mathbf{X}_{N+M,i}}(\{\mathbf{m} \in A^{N+M}; \ m_1 = b_1, \dots, m_N = b_N\})$$

$$= \sum_{\mathbf{b}\in B} \mathbb{P}(\{f \in A^{\mathbb{Z}}; f(i) = b_1, \dots, f(i+N-1) = b_N\})$$

$$= \mathbb{P}_{\mathbf{X}_{N,i}}(B).$$

Thus, with Theorem 7.5 for each $i \in \mathbb{Z}$:

$$\mathbb{S}_{\mathbb{P}_{\mathbf{X}_{N,i}}} \leq \mathbb{S}_{\mathbb{P}_{\mathbf{X}_{N+M,i}}}.$$

Now we show the important inequality for $i \in \mathbb{Z}$

$$\mathbb{S}_{\mathbb{P}_{\mathbf{X}_{N+M,i}}} \leq \mathbb{S}_{\mathbb{P}_{\mathbf{X}_{N,i}}} + \mathbb{S}_{\mathbb{P}_{\mathbf{X}_{M,i}}}.$$

Here we assume without loss of generality that

$$\mathbb{P}_{\mathbf{X}_{N,i}}(\{\mathbf{n}\}) > 0 \quad \text{for all} \quad N \in \mathbb{N} \quad \text{and} \quad \mathbf{n} \in A^N$$

and use the notation

$$\mathbb{P}_{\mathbf{X}_{M,i+N}}^{\{\mathbf{n}\}}(\{\mathbf{m}\})$$

for the conditional probability for $\{f \in A^{\mathbb{Z}}; \mathbf{X}_{M,i+N}(f) = \mathbf{m}\}$ under the condition $\{f \in A^{\mathbb{Z}}; \mathbf{X}_{N,i}(f) = \mathbf{n}\}$:

$$S_{\mathbb{P}_{\mathbf{X}_{N+M,i}}} = -\sum_{\substack{\mathbf{n}\in A^N \\ \mathbf{m}\in A^M}} \mathbb{P}_{\mathbf{X}_{N+M,i}}(\{(\mathbf{n},\mathbf{m})\})\, \mathrm{ld}\left(\mathbb{P}_{\mathbf{X}_{N+M,i}}(\{(\mathbf{n},\mathbf{m})\})\right)$$

$$= -\sum_{\substack{\mathbf{n}\in A^N \\ \mathbf{m}\in A^M}} \mathbb{P}^{\{\mathbf{n}\}}_{\mathbf{X}_{M,i+N}}(\{\mathbf{m}\})\mathbb{P}_{\mathbf{X}_{N,i}}(\{\mathbf{n}\})\, \mathrm{ld}\left(\mathbb{P}^{\{\mathbf{n}\}}_{\mathbf{X}_{M,i+N}}(\{\mathbf{m}\})\mathbb{P}_{\mathbf{X}_{N,i}}(\{\mathbf{n}\})\right)$$

$$= -\sum_{\substack{\mathbf{n}\in A^N \\ \mathbf{m}\in A^M}} \mathbb{P}^{\{\mathbf{n}\}}_{\mathbf{X}_{M,i+N}}(\{\mathbf{m}\})\mathbb{P}_{\mathbf{X}_{N,i}}(\{\mathbf{n}\})\, \mathrm{ld}\left(\mathbb{P}_{\mathbf{X}_{N,i}}(\{\mathbf{n}\})\right)$$

$$\quad - \sum_{\substack{\mathbf{n}\in A^N \\ \mathbf{m}\in A^M}} \mathbb{P}^{\{\mathbf{n}\}}_{\mathbf{X}_{M,i+N}}(\{\mathbf{m}\})\mathbb{P}_{\mathbf{X}_{N,i}}(\{\mathbf{n}\})\, \mathrm{ld}\left(\mathbb{P}^{\{\mathbf{n}\}}_{\mathbf{X}_{M,i+N}}(\{\mathbf{m}\})\right)$$

$$= -\sum_{\mathbf{n}\in A^N}\left(\mathbb{P}_{\mathbf{X}_{N,i}}(\{\mathbf{n}\})\,\mathrm{ld}\left(\mathbb{P}_{\mathbf{X}_{N,i}}(\{\mathbf{n}\})\right)\underbrace{\sum_{\mathbf{m}\in A^M}\mathbb{P}^{\{\mathbf{n}\}}_{\mathbf{X}_{M,i+N}}(\{\mathbf{m}\})}_{=1}\right)$$

$$\quad - \sum_{\substack{\mathbf{n}\in A^N \\ \mathbf{m}\in A^M}} \mathbb{P}_{\mathbf{X}_{N,i}}(\{\mathbf{n}\})\mathbb{P}^{\{\mathbf{n}\}}_{\mathbf{X}_{M,i+N}}(\{\mathbf{m}\})\, \mathrm{ld}\left(\mathbb{P}^{\{\mathbf{n}\}}_{\mathbf{X}_{M,i+N}}(\{\mathbf{m}\})\right)$$

$$= S_{\mathbb{P}_{\mathbf{X}_{N,i}}} - \sum_{\mathbf{m}\in A^M}\sum_{\mathbf{n}\in A^N} \mathbb{P}_{\mathbf{X}_{N,i}}(\{\mathbf{n}\})\mathbb{P}^{\{\mathbf{n}\}}_{\mathbf{X}_{M,i+N}}(\{\mathbf{m}\})\, \mathrm{ld}\left(\mathbb{P}^{\{\mathbf{n}\}}_{\mathbf{X}_{M,i+N}}(\{\mathbf{m}\})\right).$$

Now we examine the term

$$\sum_{\mathbf{n}\in A^N} \mathbb{P}_{\mathbf{X}_{N,i}}(\{\mathbf{n}\})\mathbb{P}^{\{\mathbf{n}\}}_{\mathbf{X}_{M,i+N}}(\{\mathbf{m}\})\, \mathrm{ld}\left(\mathbb{P}^{\{\mathbf{n}\}}_{\mathbf{X}_{M,i+N}}(\{\mathbf{m}\})\right)$$

in more detail. If we consider the function

$$f : [0,1] \to \mathbb{R}, \quad x \mapsto \begin{cases} 0 & \text{if } x = 0 \\ x\,\mathrm{ld}(x) & \text{if } x \neq 0 \end{cases},$$

then f is strictly convex. With Jensen's inequality it follows:

$$f\left(\sum_{j=1}^{k} p_j x_j\right) \leq \sum_{j=1}^{k} p_j f(x_j), \quad x_1,\ldots,x_k \in [0,1],\ p_1,\ldots,p_k \geq 0,\ \sum_{j=1}^{k} p_j = 1.$$

Now we set:

$$k = |A^N|$$
$$p_j = \mathbb{P}_{\mathbf{X}_{N,i}}(\{\mathbf{n}_j\}), \quad \mathbf{n}_j \in A^N,$$
$$x_j = \mathbb{P}^{\{\mathbf{n}_j\}}_{\mathbf{X}_{M,i+N}}(\{\mathbf{m}\}), \quad \mathbf{n}_j \in A^N,$$

and we obtain:

$$\sum_{\mathbf{n}\in A^N} \mathbb{P}_{\mathbf{X}_{N,i}}(\{\mathbf{n}\})\mathbb{P}_{\mathbf{X}_{M,i+N}}^{\{\mathbf{n}\}}(\{\mathbf{m}\})\,\mathrm{ld}\left(\mathbb{P}_{\mathbf{X}_{M,i+N}}^{\{\mathbf{n}\}}(\{\mathbf{m}\})\right)$$

$$\geq \sum_{\mathbf{n}\in A^N} \mathbb{P}_{\mathbf{X}_{N,i}}(\{\mathbf{n}\})\mathbb{P}_{\mathbf{X}_{M,i+N}}^{\{\mathbf{n}\}}(\{\mathbf{m}\})\,\mathrm{ld}\left(\sum_{\mathbf{n}\in A^N}\mathbb{P}_{\mathbf{X}_{N,i}}(\{\mathbf{n}\})\mathbb{P}_{\mathbf{X}_{M,i+N}}^{\{\mathbf{n}\}}(\{\mathbf{m}\})\right)$$

$$= \mathbb{P}_{\mathbf{X}_{M,i+N}}(\{\mathbf{m}\})\,\mathrm{ld}\left(\mathbb{P}_{\mathbf{X}_{M,i+N}}(\{\mathbf{m}\})\right).$$

Substituting above results in:

$$\mathbb{S}_{\mathbb{P}_{\mathbf{X}_{N+M,i}}} \leq \mathbb{S}_{\mathbb{P}_{\mathbf{X}_{N,i}}} - \sum_{\mathbf{m}\in A^M} \mathbb{P}_{\mathbf{X}_{M,i+N}}(\{\mathbf{m}\})\,\mathrm{ld}\left(\mathbb{P}_{\mathbf{X}_{M,i+N}}(\{\mathbf{m}\})\right)$$

$$= \mathbb{S}_{\mathbb{P}_{\mathbf{X}_{N,i}}} + \mathbb{S}_{\mathbb{P}_{\mathbf{X}_{M,i+N}}}$$

$$= \mathbb{S}_{\mathbb{P}_{\mathbf{X}_{N,i}}} + \mathbb{S}_{\mathbb{P}_{\mathbf{X}_{M,i}}}.$$

In summary, for each $i \in \mathbb{Z}$ and $N, M \in \mathbb{N}$ the double inequality

$$\mathbb{S}_{\mathbb{P}_{\mathbf{X}_{N,i}}} \leq \mathbb{S}_{\mathbb{P}_{\mathbf{X}_{N+M,i}}} \leq \mathbb{S}_{\mathbb{P}_{\mathbf{X}_{N,i}}} + \mathbb{S}_{\mathbb{P}_{\mathbf{X}_{M,i}}}$$

is available. Since the information source is stationary and the character set A is finite,

$$\mathbb{S}_{\mathbb{P}_{\mathbf{X}_{N,i}}} = \mathbb{S}_{\mathbb{P}_{\mathbf{X}_{N,j}}} < \infty \quad \text{for all} \quad N \in \mathbb{N} \text{ and } i,j \in \mathbb{Z}.$$

applies. Therefore, we simplify to

$$\mathbb{S}_{\mathbb{P}_{\mathbf{X}_N}} \leq \mathbb{S}_{\mathbb{P}_{\mathbf{X}_{N+M}}} \leq \mathbb{S}_{\mathbb{P}_{\mathbf{X}_N}} + \mathbb{S}_{\mathbb{P}_{\mathbf{X}_M}}.$$

From this it follows for $Q \in \mathbb{N}$:

$$\mathbb{S}_{\mathbb{P}_{\mathbf{X}_Q}} \leq Q\,\mathbb{S}_{\mathbb{P}_{\mathbf{X}_1}}$$

and thus the existence of $h \geq 0$ with

$$h := \liminf_{Q\to\infty} \frac{\mathbb{S}_{\mathbb{P}_{\mathbf{X}_Q}}}{Q}.$$

It remains to show that h is the only accumulation point of the sequence $\left\{\dfrac{\mathbb{S}_{\mathbb{P}_{\mathbf{X}_Q}}}{Q}\right\}$, $Q \in \mathbb{N}$.

If $N' \in \mathbb{N}$ is chosen such that for a given $\epsilon > 0$ the following applies:

$$\frac{\mathbb{S}_{\mathbb{P}_{\mathbf{X}_{N'}}}}{N'} < h + \epsilon,$$

then for every $N \in \mathbb{N}$, $N > N'$ there is a $S \in \mathbb{N}$, $S > 1$, with

$$(S-1)N' < N \leq SN'.$$

Because of

$$\mathbb{S}_{\mathbb{P}_{\mathbf{X}_N}} \leq \mathbb{S}_{\mathbb{P}_{\mathbf{X}_{SN'}}} \leq S\mathbb{S}_{\mathbb{P}_{\mathbf{X}_{N'}}}$$

it follows:

$$\frac{\mathbb{S}_{\mathbb{P}_{\mathbf{X}_N}}}{N} \leq \frac{S}{N}\mathbb{S}_{\mathbb{P}_{\mathbf{X}_{N'}}} < \frac{S}{(S-1)N'}\mathbb{S}_{\mathbb{P}_{\mathbf{X}_{N'}}} < \frac{S}{S-1}(h+\epsilon) \overset{N\to\infty}{\longrightarrow} h+\epsilon,$$

since with $N \to \infty$ also $S \to \infty$.

Analogous to Sect. 5.2, one can connect an information source to a channel (given by conditional probabilities) and characterize the transmission through the transinformation or channel capacity (see [HeHo74] for this).

Density Functions and Entropy

9.1   Integration

Probability measures on the Borel σ-field $\mathcal{B}^n$ are often represented by density functions. For this representation, one needs an integration theory, which we now recapitulate (see [Bau92]). With $\bar{\mathbb{R}} := \mathbb{R} \cup \{-\infty, +\infty\}$ an extension of the set of all real numbers is defined. The algebraic structure of $\mathbb{R}$ is extended to $\bar{\mathbb{R}}$ as follows: For all $a \in \mathbb{R}$ the following applies:

$$a + (\pm\infty) = (\pm\infty) + a = (\pm\infty) + (\pm\infty) = (\pm\infty), \quad +\infty - (-\infty) = +\infty,$$

$$a \cdot (\pm\infty) = (\pm\infty) \cdot a = \begin{cases} (\pm\infty), & \text{for } a > 0, \\ 0, & \text{for } a = 0, \\ (\mp\infty), & \text{for } a < 0, \end{cases}$$

$$(\pm\infty) \cdot (\pm\infty) = +\infty, \quad (\pm\infty) \cdot (\mp\infty) = -\infty, \quad \frac{a}{\pm\infty} = 0.$$

Thus, $\bar{\mathbb{R}}$ is not a field. The signs at $\pm\infty$ must not be combined in the above formulas, because the expressions "$+\infty + (-\infty)$" and "$-\infty + (+\infty)$" are not defined. Caution is advised with the limit theorems:

$$\lim_{x \to +\infty} \left(x \cdot \frac{1}{x} \right) \neq (+\infty) \cdot 0 = 0.$$

If one supplements the order structure of $\mathbb{R}$ by $-\infty < a$, $a < +\infty$ for all $a \in \mathbb{R}$ and $-\infty < +\infty$, then $(\bar{\mathbb{R}}, \leq)$ is an ordered set. Due to topological considerations, we can agree, without the corresponding limit theorems, that the sequence $\{n\}_{n \in \mathbb{N}}$ has the limit $+\infty \in \bar{\mathbb{R}}$ possesses. For "$+\infty$" we often write "∞".

The basis of the integration theory to be developed now is the concept of measure.

© The Author(s), under exclusive license to Springer-Verlag GmbH, DE, part of
Springer Nature 2024
S. Schäffler, *Mathematics of Information*, Mathematics Study Resources 9,
https://doi.org/10.1007/978-3-662-69102-1_9

Definition 9.1 (Measure, Measure Space) Let $(\Omega, \mathcal{S})$ be a measurable space, then a mapping

(M1) $\mu : \mathcal{S} \to [0, \infty]$,
(M2) $\mu(\emptyset) = 0$,
(M3) for each sequence $\{A_i\}_{i \in \mathbb{N}}$ of pairwise disjoint sets with $A_i \in \mathcal{S}$, $i \in \mathbb{N}$, applies:

$$\mu\left(\bigcup_{i=1}^{\infty} A_i\right) = \sum_{i=1}^{\infty} \mu(A_i),$$

is referred to as **measure**. The triple $(\Omega, \mathcal{S}, \mu)$ is called **measure space**.   ◁

A probability measure $\mathbb{P}$ is always a special measure with $\mathbb{P}(\Omega) = 1$. Usually, in integration theory, one starts with the integration of simple functions.

Definition 9.2 (elementary functions) Let $(\Omega, \mathcal{S})$ be a measurable space. A $\mathcal{S}$-$\mathcal{B}$ -measurable function

$$e : \Omega \to \mathbb{R}$$

is called an **elementary function** if it only takes a finite number of different function values.   ◁

A special elementary function is the indicator function

$$I_A : \Omega \to \mathbb{R}, \quad \omega \mapsto \begin{cases} 1 & \text{if } \omega \in A \\ 0 & \text{else} \end{cases},$$

which indicates whether ω is an element of a set $A \in \mathcal{S}$. Elementary functions can be represented using indicator functions.

▶   **Theorem 9.3 (Representation of elementary functions)** *Let $(\Omega, \mathcal{S})$ be a measurable space. If*

is an elementary function, then there exist a natural number n, *pairwise disjoint*
$$e : \Omega \to \mathbb{R}$$

sets $A_1, \ldots, A_n \in \mathcal{S}$ and real numbers $\alpha_1, \ldots, \alpha_n$ with:

$$e = \sum_{i=1}^{n} \alpha_i I_{A_i}, \quad \bigcup_{i=1}^{n} A_i = \Omega.$$

Since e only takes on a finite number of different function values, we choose n to be equal to the number of different function values of e and we set $\alpha_1, \ldots, \alpha_n$ equal to the different function values. Since

$$\{\alpha_i\} \in \mathcal{B} \quad \text{for} \quad i = 1, \ldots, n$$

and since e $\mathcal{S}$-$\mathcal{B}$-measurable, it follows that

$$\{\omega \in \Omega; \ e(\omega) = \alpha_i\} =: e^{-1}(\{\alpha_i\}) \in \mathcal{S} \quad \text{for all} \quad i = 1, \ldots, n.$$

Furthermore, it holds that:

$$e^{-1}(\{\alpha_i\}) \cap e^{-1}(\{\alpha_j\}) = \emptyset \quad \text{for } i \neq j$$

and

$$\bigcup_{i=1}^{n} e^{-1}(\{\alpha_i\}) = \Omega.$$

With

$$A_i := e^{-1}(\{\alpha_i\}), \ i = 1, \ldots, n,$$

the theorem is proven. **q.e.d.**

The representation of e considered in Theorem 9.3 is called a **normal representation** of e. The normal representation chosen in the proof of Theorem 9.3 is called the **shortest normal representation** of e, since all $\alpha_i, i = 1, \ldots, n$, are assumed to be pairwise different.

The sum, difference, and product of elementary functions are elementary functions. For all $c \in \mathbb{R}$, $c \cdot e$ is also an elementary function, if e is an elementary function.

Now we consider non-negative elementary functions on a measure space $(\Omega, \mathcal{S}, \mu)$ and define the (μ-) integral of these functions.

Definition 9.4 ((μ-) Integral of non-negative elementary functions) Let $(\Omega, \mathcal{S}, \mu)$ be a measure space and $e : \Omega \to \mathbb{R}_0^+$,

$$e = \sum_{i=1}^{n} \alpha_i I_{A_i}, \quad \alpha_i \geq 0, \quad i = 1, \ldots, n,$$

a non-negative elementary function in normal form, then

$$\int e \, d\mu := \int_{\Omega} e \, d\mu := \sum_{i=1}^{n} \alpha_i \cdot \mu(A_i)$$

is referred to as the (μ-)Integral of e over Ω. ◁

In order for $\int e \, d\mu$ to be well-defined, it is naturally necessary to show that $\int e \, d\mu$ is independent of the choice of normal form for e.

▶   **Theorem 9.5 (Independence of the integral from the normal form)** *Let* $(\Omega, \mathcal{S}, \mu)$ *be a measure space and* $e : \Omega \to \mathbb{R}_0^+$ *a non-negative elementary function with the normal forms*

$$e = \sum_{i=1}^{n} \alpha_i I_{A_i} = \sum_{j=1}^{m} \beta_j I_{B_j}, \quad \alpha_i, \beta_j \geq 0, \ i = 1, \ldots, n, \ j = 1, \ldots, m,$$

then

$$\sum_{i=1}^{n} \alpha_i \cdot \mu(A_i) = \sum_{j=1}^{m} \beta_j \cdot \mu(B_j).$$

applies.◁

Since

$$\Omega = \bigcup_{i=1}^{n} A_i = \bigcup_{j=1}^{m} B_j,$$

applies

$$A_i = \bigcup_{j=1}^{m} (A_i \cap B_j), \ i = 1, \ldots, n \text{ and } B_j = \bigcup_{i=1}^{n} (B_j \cap A_i), \ j = 1, \ldots, m.$$

Thus we obtain

$$\mu(A_i) = \sum_{j=1}^{m} \mu(A_i \cap B_j), \ i = 1, \ldots, n, \text{ and}$$

$$\mu(B_j) = \sum_{i=1}^{n} \mu(B_j \cap A_i), \ j = 1, \ldots, m.$$

So it follows:

$$\sum_{i=1}^{n} \alpha_i \mu(A_i) = \sum_{i=1}^{n} \alpha_i \sum_{j=1}^{m} \mu(A_i \cap B_j) = \sum_{\substack{i \in \{1,\ldots,n\} \\ j \in \{1,\ldots,m\}}} \alpha_i \mu(A_i \cap B_j),$$

$$\sum_{j=1}^{m} \beta_j \mu(B_j) = \sum_{j=1}^{m} \beta_j \sum_{i=1}^{n} \mu(B_j \cap A_i) = \sum_{\substack{i \in \{1,\ldots,n\} \\ j \in \{1,\ldots,m\}}} \beta_j \mu(A_i \cap B_j).$$

q.e.d.

Let now $(A_i \cap B_j) \neq \emptyset$ for a pair of indices (i,j), so is $\alpha_i = e(\omega) = \beta_j$ for all $\omega \in (A_i \cap B_j)$. Thus, $\alpha_i = \beta_j$ for all pairs of indices (i,j) with $\mu(A_i \cap B_j) \neq 0$. We obtain

$$\int e\, d\mu = \sum_{i=1}^{n} \alpha_i \mu(A_i) = \sum_{\substack{i \in \{1,\dots,n\} \\ j \in \{1,\dots,m\}}} \alpha_i \mu(A_i \cap B_j) = \sum_{j=1}^{m} \beta_j \mu(B_j).$$

Let E be the set of all non-negative elementary functions with respect to a measure space $(\Omega, \mathcal{S}, \mu)$, we obtain a mapping

$$\text{Int}: E \to \mathbb{R}_0^+ \cup \{\infty\}, \ e \mapsto \int e\, d\mu.$$

The following properties of Int can be easily proven:

(i) $\int I_A d\mu = \mu(A)$ for all $A \in \mathcal{S}$.
(ii) $\int (\alpha e) d\mu = \alpha \int e\, d\mu$ for all $e \in E$, $\alpha \in \mathbb{R}_0^+$.
(iii) $\int (u + v) d\mu = \int u\, d\mu + \int v\, d\mu$ for all $u, v \in E$.
(iv) If $u(\omega) \leq v(\omega)$ for all $\omega \in \Omega$, then

$$\int u\, d\mu \leq \int v\, d\mu \quad \text{for all} \quad u, v \in E.$$

Considering the set $\bar{\mathbb{R}}$, the set

$$\bar{\mathcal{B}} := \{A \in \mathcal{P}(\bar{\mathbb{R}}); \ A \cap \mathbb{R} \in \mathcal{B}\}$$

forms a σ-field over $\bar{\mathbb{R}}$. To extend the concept of integration to a larger class of functions, we need the following definition.

Definition 9.6 (numerical function) A function defined on a non-empty set $A \subseteq \Omega\, f : A \to \bar{\mathbb{R}}$ is called a **numerical function**. ◁

Starting from a measure space $(\Omega, \mathcal{S}, \mu)$ we now want to define the $(\mu\text{-})$ integral for $\mathcal{S}\text{-}\bar{\mathcal{B}}$-measurable numerical functions. For this, we consider pointwise convergence and the monotonicity of sequences of functions.

Definition 9.7 (Pointwise Convergence and Monotonicity of Sequences of Functions) Let $\{f_n\}_{n \in \mathbb{N}}$ be a sequence of functions

$$f_n : \Omega \to \bar{\mathbb{R}}, \quad n \in \mathbb{N}.$$

$\{f_n\}_{n \in \mathbb{N}}$ is called **pointwise convergent**, if there is a function $f : \Omega \to \bar{\mathbb{R}}$ with

$$\lim_{n \to \infty} f_n(\omega) = f(\omega) \quad \text{for all} \quad \omega \in \Omega.$$

$\{f_n\}_{n \in \mathbb{N}}$ is called **monotonically increasing**, if

$$f_n(\omega) \leq f_{n+1}(\omega) \quad \text{for all} \quad \omega \in \Omega,\ n \in \mathbb{N}.$$

$\{f_n\}_{n \in \mathbb{N}}$ is called **monotonically decreasing**, if

$$f_n(\omega) \geq f_{n+1}(\omega) \quad \text{for all} \quad \omega \in \Omega, \; n \in \mathbb{N}. \qquad \triangleleft$$

If a monotonically increasing sequence $\{f_n\}_{n\in\mathbb{N}}$ of functions converges pointwise to f, this is denoted by $f_n \uparrow f$ (for a monotonically decreasing sequence: $f_n \downarrow f$).

We now examine with $\bar{\mathbb{R}}_0^+ := \{x \in \mathbb{R};\; x \geq 0\} \cup \{\infty\}$ non-negative numerical functions that are given as the limit of a sequence of elementary functions.

▶ **Theorem 9.8 (Limit values of special sequences of elementary functions)** *Let* $(\Omega, \mathcal{S})$ *be a measurable space and* $f : \Omega \to \bar{\mathbb{R}}_0^+$ *a non-negative, $\mathcal{S}$-$\bar{\mathcal{B}}$-measurable numerical function, then there exists a monotonically increasing sequence* $\{e_n\}_{n\in\mathbb{N}}$ *of non-negative elementary functions*

$$e_n : \Omega \to \mathbb{R}_0^+, \quad n \in \mathbb{N},$$

with $e_n \uparrow f.\triangleleft$

We consider for $n \in \mathbb{N}$ the non-negative functions

$$e_n : \Omega \to \mathbb{R}_0^+,$$

$$\omega \mapsto \sum_{k=1}^{n \cdot 2^n} \frac{k-1}{2^n} \cdot I_{\{\omega \in \Omega;\; \frac{k-1}{2^n} \leq f(\omega) < \frac{k}{2^n}\}}(\omega) + n \cdot I_{\{\omega \in \Omega;\; f(\omega) \geq n\}}(\omega).$$

Since the intervals $(-\infty, c)$, $c \in \mathbb{R}$, all lie in $\bar{\mathcal{B}}$, it follows from the $\mathcal{S}$-$\bar{\mathcal{B}}$-measurability of f, that with $a, b \in \mathbb{R}$, $a < b$ the sets

$$\{\omega \in \Omega;\; a \leq f(\omega) < b\}$$
$$= \{\omega \in \Omega;\; -\infty < f(\omega) < a\}^c \cap \{\omega \in \Omega;\; -\infty < f(\omega) < b\}$$

lie in $\mathcal{S}$. Thus, the functions e_n are elementary for all $n \in \mathbb{N}$.

For

$$\bar{\omega} \in \left\{ \omega \in \Omega;\; \frac{k-1}{2^n} \leq f(\omega) < \frac{k}{2^n} \right\}, \; k = 1, \dots, n \cdot 2^n, \; n \in \mathbb{N},$$

applies:

$$e_{n+1}(\bar{\omega}) \geq \frac{2(k-1)}{2^{n+1}} = \frac{k-1}{2^n} = e_n(\bar{\omega}),$$

since

$$\bar{\omega} \in \left\{ \omega \in \Omega;\; \frac{2(k-1)}{2^{n+1}} \leq f(\omega) < \frac{2k}{2^{n+1}} \right\}.$$

Is $f(\bar{\omega}) \geq n$, so is also $e_{n+1}(\bar{\omega}) \geq n = e_n(\bar{\omega})$. Thus, $\{e_n\}_{n\in\mathbb{N}}$ is monotonically increasing. Now let $f(\bar{\omega}) < \infty$ be, so for $f(\bar{\omega}) < n, \; n \in \mathbb{N}$:

$$0 \leq f(\bar{\omega}) - e_n(\bar{\omega}) < \frac{1}{2^n}$$

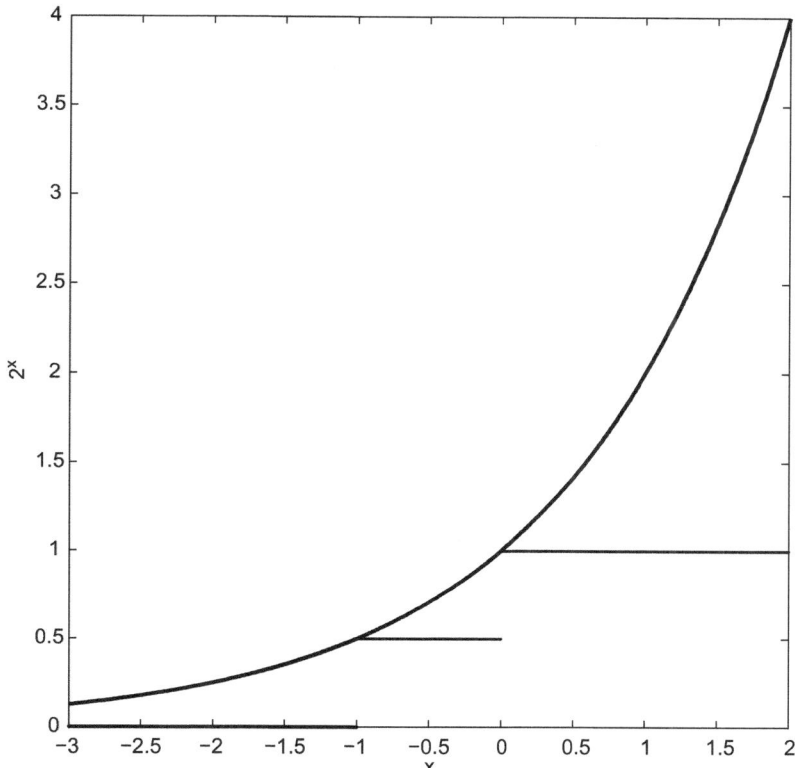

Fig. 9.1 Approximation of $x \mapsto 2^x$ with e_1

and thus

$$e_n(\bar{\omega}) \uparrow f(\bar{\omega}).$$

Is $f(\bar{\omega}) = \infty$, so is $e_n(\bar{\omega}) = n \uparrow \infty$. **q.e.d.**

It is important to note that in proving Theorem 9.8 not only the existence of $\{e_n\}_{n\in\mathbb{N}}$ is shown, but for a given f the sequence $\{e_n\}_{n\in\mathbb{N}}$ can be explicitly specified. The following Fig. 9.1 shows the approximation of the function

$$f : \mathbb{R} \to \mathbb{R}_0^+, \quad x \mapsto 2^x$$

by

$$e_1 : \mathbb{R} \to \mathbb{R}_0^+,$$

$$x \mapsto \sum_{k=1}^{2} \frac{k-1}{2} \cdot I_{\{x\in\mathbb{R};\ \frac{k-1}{2}\leq 2^x < \frac{k}{2}\}}(x) + I_{\{x\in\mathbb{R};\ 2^x\geq 1\}}(x).$$

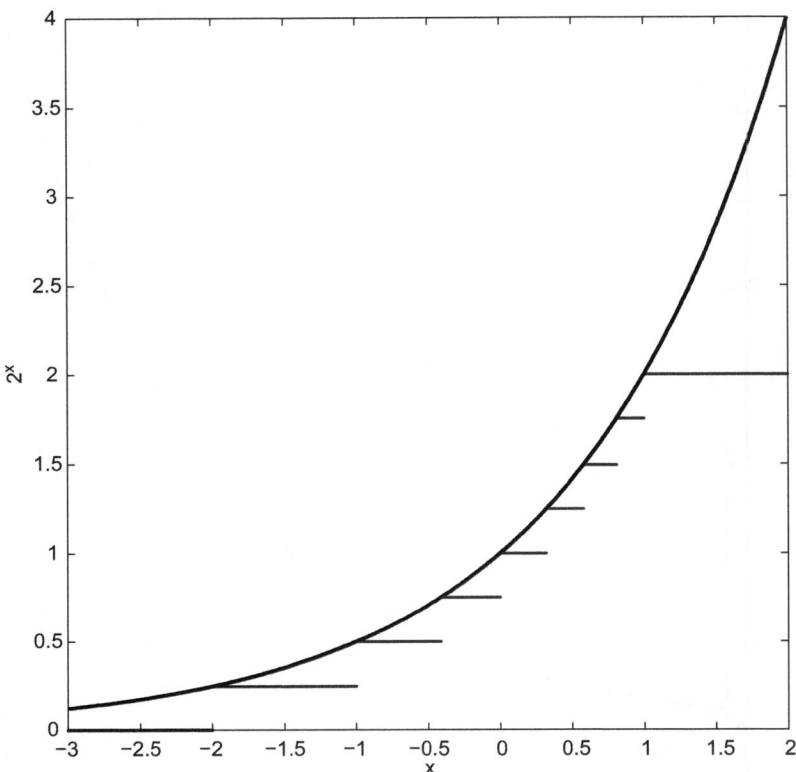

Fig. 9.2 Approximation of $x \mapsto 2^x$ with e_2

in the interval $x \in [-3, 2]$, while Fig. 9.2 shows the approximation of the function

$$f : \mathbb{R} \to \mathbb{R}_0^+, \quad x \mapsto 2^x$$

by

$$e_2 : \mathbb{R} \to \mathbb{R}_0^+,$$

$$x \mapsto \sum_{k=1}^{8} \frac{k-1}{4} \cdot I_{\left\{x \in \mathbb{R}; \ \frac{k-1}{4} \leq 2^x < \frac{k}{4}\right\}}(x) + 2 \cdot I_{\left\{x \in \mathbb{R}; \ 2^x \geq 2\right\}}(x)$$

also in the interval $x \in [-3, 2]$.

Figure 9.3 shows the approximation of the function

$$f : \mathbb{R} \to \mathbb{R}_0^+, \quad x \mapsto 3e^{-x^2}$$

by

$$e_2 : \mathbb{R} \to \mathbb{R}_0^+,$$

$$x \mapsto \sum_{k=1}^{8} \frac{k-1}{4} \cdot I_{\left\{x \in \mathbb{R}; \ \frac{k-1}{4} \leq 3e^{-x^2} < \frac{k}{4}\right\}}(x) + 2 \cdot I_{\left\{x \in \mathbb{R}; \ 3e^{-x^2} \geq 2\right\}}(x)$$

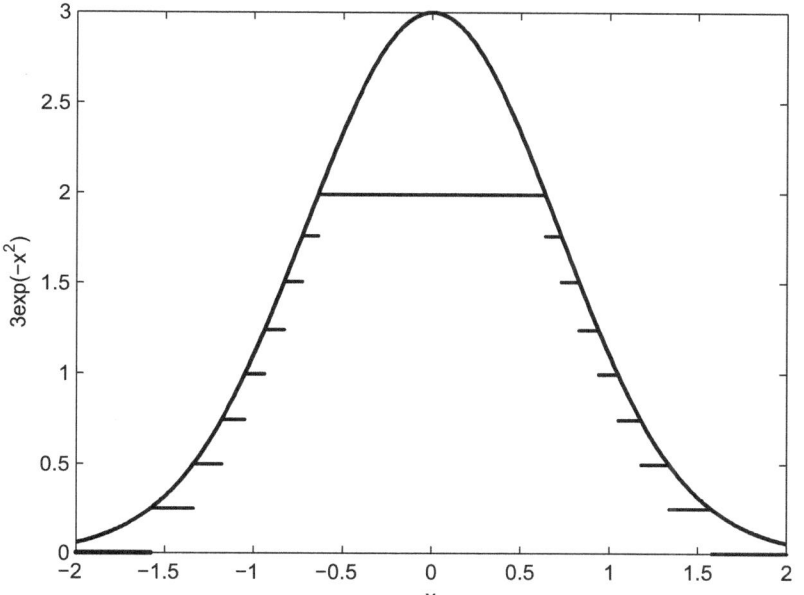

Fig. 9.3 Approximation of $x \mapsto 3e^{-x^2}$ with e_2

in the interval $x \in [-2, 2]$.

After these preparations, we are able to extend the $(\mu\text{-})$ integration to a special class of functions in an obvious way.

Definition 9.9 ((μ-) Integral for $\mathcal{S}$-$\bar{\mathcal{B}}$-measurable, non-negative numerical functions) Let $(\Omega, \mathcal{S}, \mu)$ be a measure space and $f : \Omega \to \bar{\mathbb{R}}_0^+$ a $\mathcal{S}$-$\bar{\mathcal{B}}$-measurable, non-negative numerical function. Furthermore, let $\{e_n\}_{n \in \mathbb{N}}$ be a monotonically increasing sequence of non-negative elementary functions

$$e_n : \Omega \to \mathbb{R}_0^+, \quad n \in \mathbb{N} \text{ with } e_n \uparrow f,$$

we define by

$$\int f d\mu := \int_\Omega f d\mu := \lim_{n \to \infty} \int e_n d\mu$$

the $(\mu\text{-})$ integral of f over Ω. ◁

Since the sequence $\{e_n\}_{n \in \mathbb{N}}$ considered in Definition 9.1 is not uniquely determined by $e_n \uparrow f$, the $(\mu\text{-})$ integral of f is well-defined only if $\int f d\mu$ is independent of the choice of the sequence $\{e_n\}_{n \in \mathbb{N}}$ of non-negative elementary functions with $e_n \uparrow f$. Therefore, the following theorem is of crucial importance.

▶ **Theorem 9.10 ($\int f d\mu$ is well-defined)** *Let $(\Omega, \mathcal{S}, \mu)$ be a measure space and $f : \Omega \to \bar{\mathbb{R}}_0^+$ a $\mathcal{S}$-$\bar{\mathcal{B}}$-measurable, non-negative numerical function, then for two monotonically increasing sequences $\{e_n\}_{n\in\mathbb{N}}$ and $\{h_n\}_{n\in\mathbb{N}}$ of non-negative elementary functions with $e_n \uparrow f$ and $h_n \uparrow f$:*

$$\lim_{n\to\infty} \int e_n d\mu = \lim_{n\to\infty} \int h_n d\mu.$$

◁

To prove this theorem, we use the following lemma.

Lemma 9.10 (Inequalities for elementary functions and their integrals) *Let $(\Omega, \mathcal{S}, \mu)$ be a measure space, $e : \Omega \to \mathbb{R}_0^+$ a non-negative elementary function and $f : \Omega \to \bar{\mathbb{R}}_0^+$ a $\mathcal{S}$-$\bar{\mathcal{B}}$-measurable, non-negative numerical function with $e(\omega) \le f(\omega)$ for all $\omega \in \Omega$. Let further $\{f_n\}_{n\in\mathbb{N}}$ be a monotonically increasing sequence of non-negative elementary functions with $f_n \uparrow f$, then:*

$$\int e d\mu \le \lim_{n\to\infty} \int f_n d\mu.$$

◁

Let $e = \sum\limits_{i=1}^p c_i I_{E_i}$. We consider the following sets:

$$T := \{\omega \in \Omega;\ e(\omega) > 0\} \text{ and } A_n^\varepsilon := \{\omega \in \Omega;\ f_n(\omega) + \varepsilon > e(\omega)\}$$

for all $\varepsilon > 0$ und $n \in \mathbb{N}$.

1. Case: $\int e d\mu = +\infty$, then there exists a $j \in \{1, \ldots, p\}$ with $\mu(E_j) = +\infty$ and $c_j > 0$.

In addition to

$$A_k^\varepsilon \cap E_j \subseteq A_{k+1}^\varepsilon \cap E_j \quad \text{for all} \quad k \in \mathbb{N},$$

$$\bigcup_{k=1}^\infty (A_k^\varepsilon \cap E_j) = E_j.$$

also applies. Now let

$$C_1 := A_1^\varepsilon \cap E_j,$$
$$C_m := (A_m^\varepsilon \cap E_j) \setminus (C_{m-1} \cup \ldots \cup C_1) \quad \text{for all} \quad m \in \mathbb{N}, m \ge 2,,$$

we obtain:

$$\lim_{k\to\infty} \mu\left(A_k^\varepsilon \cap E_j\right) = \lim_{k\to\infty} \mu\left(\bigcup_{m=1}^k C_m\right) = \lim_{k\to\infty} \sum_{m=1}^k \mu(C_m)$$
$$= \sum_{m=1}^\infty \mu(C_m) = \mu\left(\bigcup_{m=1}^\infty C_m\right) = \mu(E_j)$$
$$= +\infty.$$

Since

$$f_n(\omega) > e(\omega) - \varepsilon \quad \text{for all} \quad \omega \in A_n^\varepsilon, \ n \in \mathbb{N},$$

applies for $0 < \varepsilon < c_j$:

$$\int f_n d\mu \geq \int f_n I_{A_n^\varepsilon \cap E_j} d\mu \geq \int (e - \varepsilon) I_{A_n^\varepsilon \cap E_j} d\mu$$

$$\geq (c_j - \varepsilon) \mu\left(A_n^\varepsilon \cap E_j\right) \overset{n \to \infty}{\longrightarrow} +\infty.$$

Case 2: $\int e\, d\mu < +\infty$, we show that with $T := \{\omega \in \Omega; \ e(\omega) > 0\}$ for every $\varepsilon > 0$ with $0 < \varepsilon < \min\{c_i; c_i > 0\}$ there is a natural number $n_0(\varepsilon)$ such that for all $n \geq n_0(\varepsilon), n \in \mathbb{N}$, the following holds:

$$\int f_n d\mu \geq \int e \cdot I_T d\mu - \varepsilon - \varepsilon \mu(T).$$

Since $\mu(T) < \infty$ and $\int e\, d\mu = \int e \cdot I_T d\mu$, our claim is proven with the above inequality. Because of

$$T = \left(T \cap A_n^\varepsilon\right) \cup \left(T \cap (A_n^\varepsilon)^c\right)$$

it holds for all $n \in \mathbb{N}$:

$$\int f_n d\mu \geq \int f_n I_{T \cap A_n^\varepsilon} d\mu \geq \int (e - \varepsilon) I_{T \cap A_n^\varepsilon} d\mu = \int e I_{T \cap A_n^\varepsilon} d\mu - \varepsilon \mu\left(T \cap A_n^\varepsilon\right),$$

thus

$$\int f_n d\mu \geq \int e I_{T \cap A_n^\varepsilon} d\mu - \varepsilon \mu(T) = \int e I_T d\mu - \int e I_{T \cap (A_n^\varepsilon)^c} d\mu - \varepsilon \mu(T).$$

Since

$$T \cap (A_n^\varepsilon)^c = \{\omega \in \Omega; \ e(\omega) \geq f_n(\omega) + \varepsilon > 0\},$$

the existence of a $n_0(\varepsilon)$ with

$$\int e \cdot I_{T \cap (A_n^\varepsilon)^c} d\mu < \varepsilon \quad \text{for all } n \geq n_0(\varepsilon)$$

can be demonstrated.   **q.e.d.**

Now we are able to provide a very short proof for Theorem 9.10.

From Theorem 9.10 From $e_n \uparrow f$ and $h_n \uparrow f$ it follows for every fixed $k \in \mathbb{N}$, that

$$f = \lim_{n \to \infty} e_n \geq h_k.$$

Thus, according to Lemma 1

$$\lim_{n \to \infty} \int e_n d\mu \geq \int h_k d\mu \quad \text{for all } k \in \mathbb{N}$$

and with $k \to \infty$:

$$\lim_{n\to\infty} \int e_n d\mu \geq \lim_{k\to\infty} \int h_k \, d\mu.$$

But also $f = \lim_{k\to\infty} h_k \geq e_n$ for all $n \in \mathbb{N}$ applies, also follows with Lemma 1

$$\lim_{n\to\infty} \int e_n d\mu \leq \lim_{k\to\infty} \int h_k d\mu. \qquad \qquad \textbf{q.e.d.}$$

The following definition serves to extend the definition of the (μ-)integral to a larger class of functions.

Definition 9.11 (Positive part, Negative part of a numerical function) Let $(\Omega, \mathcal{S})$ be a measure space and $f : \Omega \to \bar{\mathbb{R}}$ a $\mathcal{S}$-$\bar{\mathcal{B}}$-measurable numerical function, then the function

$$f^+ : \Omega \to \bar{\mathbb{R}}_0^+, \quad \omega \mapsto \begin{cases} f(\omega) & \text{if } f(\omega) \geq 0 \\ 0 & \text{else} \end{cases}$$

is called the positive part of f and the function

$$f^- : \Omega \to \bar{\mathbb{R}}_0^+, \quad \omega \mapsto \begin{cases} -f(\omega) & \text{if } f(\omega) \leq 0 \\ 0 & \text{else} \end{cases}$$

is called the negative part of f.   ◁

The following properties of f^+ and f^- are immediately apparent (see Fig. 9.4, 9.5 and 9.6):

(i) $f^+(\omega) \geq 0$, $f^-(\omega) \geq 0$ for all $\omega \in \Omega$.
(ii) f^+ and f^- are $\mathcal{S}$-$\bar{\mathcal{B}}$-measurable numerical functions.
(iii) $f = f^+ - f^-$.

With the help of the positive and negative parts of a measurable numerical function $f : \Omega \to \bar{\mathbb{R}}$, we can extend the ($\mu$-)integral to measurable numerical functions.

Definition 9.12 ((μ-) integrable, (μ-) quasi-integrable, (μ-) Integral) Let $(\Omega, \mathcal{S}, \mu)$ be a measure space and $f : \Omega \to \bar{\mathbb{R}}$ a $\mathcal{S}$-$\bar{\mathcal{B}}$-measurable numerical function.

f is called (μ-) integrable, if $\int f^+ d\mu < \infty$ and $\int f^- d\mu < \infty$.
f is called (μ-) quasi-integrable, if $\int f^+ d\mu < \infty$ or $\int f^- d\mu < \infty$.

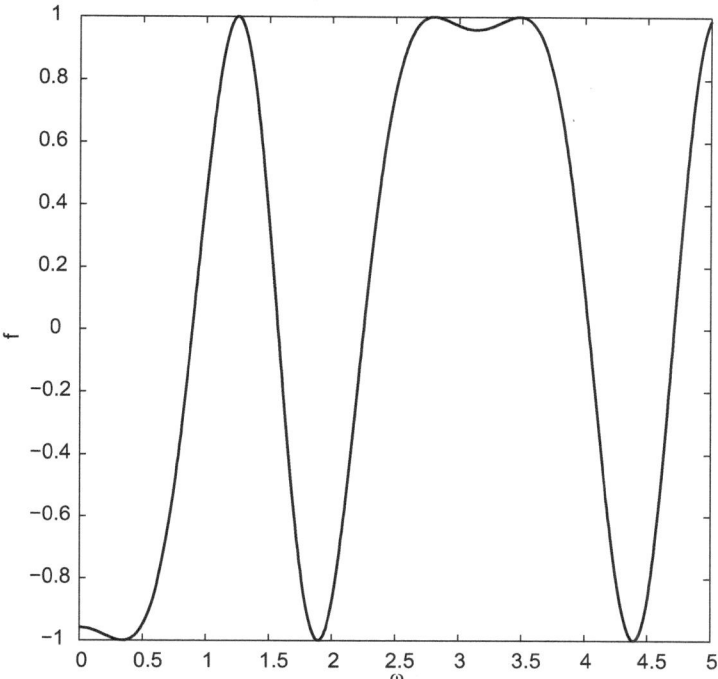

Fig. 9.4   Function f

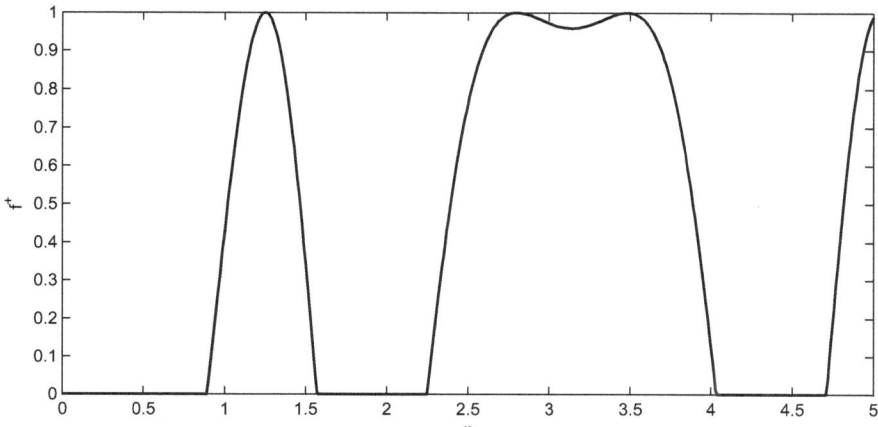

Fig. 9.5   Positive part f^+ of f

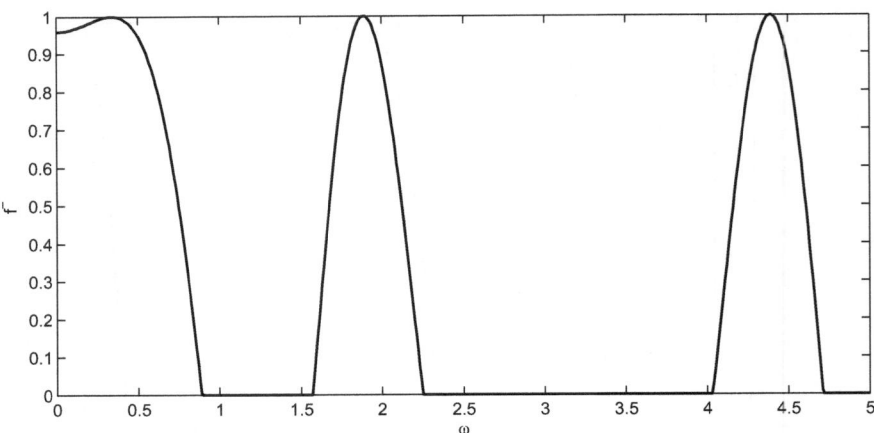

Fig. 9.6 Negative part f^- of f

If $f(\mu\text{-})$ is quasi-integrable, then by

$$\int f d\mu := \int_\Omega f d\mu := \int f^+ d\mu - \int f^- d\mu$$

the (μ-)integral of f over Ω is defined.   ◁

Due to Theorem 9.10 is $\int f d\mu$ well-defined. As (μ-)integral over a set $A \in \mathcal{S}$ we define for (μ-)quasi-integrable $f \cdot I_A$:

$$\int_A f d\mu := \int f \cdot I_A d\mu.$$

Specifically considering the measure space $(\mathbb{R}^n, \mathcal{B}^n)$, $n \in \mathbb{N}$, there is a unique measure

$$\lambda^n : \mathcal{B}^n \to [0, \infty]$$

with

$$\lambda^n((a_1, b_1] \times \ldots \times (a_n, b_n]) = \prod_{i=1}^n (b_i - a_i), \quad a_j < b_j, j = 1, \ldots, n,$$

(for proof see [Klenke05]). This measure is referred to as the **Lebesgue measure**. Furthermore, the (λ^n-) integral is referred to as the Lebesgue integral. If $f(\lambda^n\text{-})$ integrable, then f is called Lebesgue-integrable.

9.2   Densities

Let $(\Omega, \mathcal{S}, \mu)$ be a measure space and let there be a $\mathcal{S}$-$\mathcal{B}$-measurable function

$$f : \Omega \to \mathbb{R}_0^+ \quad \text{with} \quad \int f d\mu = 1$$

given, we obtain a probability measure $\mathbb{P}$ on $\mathcal{S}$ by

$$\mathbb{P}_f : \mathcal{S} \to [0, 1], \quad A \mapsto \int_A f d\mu.$$

The function f is referred to as **density** or **density function** of $\mathbb{P}_f$ with respect to μ. The most interesting case for applications refers to the measure space $(\mathbb{R}^n, \mathcal{B}^n, \lambda^n)$, $n \in \mathbb{N}$. Since every continuous function

$$f : \mathbb{R}^n \to \mathbb{R}_0^+$$

$\mathcal{B}^n$-$\mathcal{B}$-measurable is, a probability measure $\mathbb{P}_f$ on $\mathcal{B}^n$ is given by every continuous function

$$f : \mathbb{R}^n \to \mathbb{R}_0^+ \quad \text{with} \quad \int f d\lambda^n = 1.$$

If, for example, one chooses the sets

$$(a_1, b_1] \times \ldots \times (a_n, b_n] \in \mathcal{B}^n, \quad a_j < b_j, \ j = 1, \ldots, n,$$

one can calculate the corresponding probabilities by Riemann integration

$$\mathbb{P}_f((a_1, b_1] \times \ldots \times (a_n, b_n]) = \int_{a_1}^{b_1} \ldots \int_{a_n}^{b_n} f(x_1, \ldots, x_n) dx_1 \ldots dx_n.$$

Now let's consider the entropy of $(\mathbb{R}^n, \mathcal{B}^n, \mathbb{P}_f)$: For $m \in \mathbb{N}$, $m > 1$, there are intervals

$$I_1 = (-\infty, \xi_1], I_2 = (\xi_1, \xi_2], \ldots, I_m = (\xi_{m-1}, \infty)$$

with

$$\int_{I_j \times \mathbb{R}^{n-1}} f d\lambda^n = \int_{I_j}^{\infty} \int_{-\infty}^{\infty} \ldots \int_{-\infty}^{\infty} f(x_1, \ldots, x_n) dx_1 \ldots dx_n = \frac{1}{m}, \quad j = 1, \ldots, m.$$

Because of

$$-\sum_{j=1}^{m} \mathbb{P}_f(I_j \times \mathbb{R}^{n-1}) \operatorname{ld}(\mathbb{P}_f(I_j \times \mathbb{R}^{n-1})) = \operatorname{ld}(m)$$

applies

$$\mathbb{S}_{\mathbb{P}_f} \geq \mathrm{ld}(m) \quad \text{for each} \quad m \in \mathbb{N}$$

and therefore

$$\mathbb{S}_{\mathbb{P}_f} = \infty.$$

The Shannon entropy introduced in Definition 7.2 is not generally helpful for probability measures

$$\mathbb{P}_f : \mathcal{B}^n \to [0,1], \quad A \mapsto \int_A f d\lambda^n.$$

Therefore, in these cases, the differential entropy is considered.

9.3   Differential Entropy

The differential entropy to be defined now serves to compare probability measures, which are given by densities, in terms of information theory.

Definition 9.13 (Differential Entropy) Let $(\Omega, \mathcal{S}, \mu)$ be a measure space and $f : \Omega \to \mathbb{R}_0^+$ a density function with respect to μ. Let the function

$$\eta : \Omega \to \mathbb{R}, \quad \omega \mapsto f(\omega)\,\mathrm{ld}(f(\omega)) \quad \text{(remember: } 0 \cdot \mathrm{ld}(0) := 0\text{)}$$

$(\mu\text{-})$ be quasi-integrable, then

$$\mathbb{S}_f := -\int \eta d\mu$$

is referred to as the **differential entropy** of f.   ◁

The $\mathcal{S}$-$\mathcal{B}$-measurability of $\eta = f \cdot (\mathrm{ld} \circ f)$ is ensured by the $\mathcal{S}$-$\mathcal{B}$-measurability of f (density function) and by the continuity of the function ld. For $(\mathbb{R}, \mathcal{B}, \lambda)$, $a > 0$ and

$$f_a : \mathbb{R} \to \mathbb{R}_0^+, \quad x \mapsto \begin{cases} \frac{1}{a} & \text{if } 0 \leq x \leq a \\ 0 & \text{else} \end{cases}$$

it is obviously

$$\mathbb{S}_{f_a} = -\int_0^a \frac{1}{a}\,\mathrm{ld}\left(\frac{1}{a}\right) dx = \mathrm{ld}(a).$$

In contrast to Shannon entropy, the differential entropy can also take negative values.

Let now $(\Omega, \mathcal{S}, \mu)$ be a measure space and $f, g : \Omega \to \mathbb{R}_0^+$ two density functions with respect to μ. Let further the functions

$$\eta_1 : \Omega \to \mathbb{R}, \quad \omega \mapsto f(\omega)\,\mathrm{ld}(f(\omega))$$

and

$$\eta_2 : \Omega \to \mathbb{R}, \quad \omega \mapsto g(\omega) \, \mathrm{ld}(g(\omega))$$

(μ-) be quasi-integrable, then one can use the two differential entropies $\mathbb{S}_f$ and $\mathbb{S}_g$ to compare the average amount of information of the probability spaces $(\Omega, \mathcal{S}, \mathbb{P}_f)$ and $(\Omega, \mathcal{S}, \mathbb{P}_g)$. Thus, the average amount of information of the probability space $(\Omega, \mathcal{S}, \mathbb{P}_{f_2})$ with

$$\mathbb{P}_{f_2} : \mathcal{B} \to [0,1], \quad A \mapsto \int_{A \cap [0,2]} \frac{1}{2} d\lambda = \frac{1}{2} \lambda(A \cap [0,2])$$

would be greater than the average amount of information of the probability space $(\Omega, \mathcal{S}, \mathbb{P}_{f_1})$ with

$$\mathbb{P}_{f_1} : \mathcal{B} \to [0,1], \quad A \mapsto \int_{A \cap [0,1]} 1 d\lambda = \lambda(A \cap [0,1]),$$

since

$$0 = \mathbb{S}_{f_1} < \mathbb{S}_{f_2} = 1.$$

This makes sense when considering that on the one hand for $A \in \mathcal{B}$ applies:

$$\mathbb{P}_{f_1}(A) > 0 \quad \Longrightarrow \quad \mathbb{P}_{f_2}(A) > 0$$

and that on the other hand for all sets

$$M \subseteq (1, \infty) \quad \text{with} \quad M \in \mathcal{B} \quad \text{and} \quad \lambda(M \cap (1,2]) > 0$$

applies:

$$0 = \mathbb{P}_{f_1}(M) < \mathbb{P}_{f_2}(M).$$

For the Shannon entropy we get:

$$\mathbb{S}_{\mathbb{P}_{f_1}} = \mathbb{S}_{\mathbb{P}_{f_2}} = \infty.$$

In the following, we investigate the maximization of the differential entropy under constraints analogous to Chap. 4. Since the initial situation is much more complicated than with discrete systems, we do not consider the problem in full generality, but only within a framework that does not require variational calculus. For this we need the following lemma.

Lemma 9.15 (Gibbs Inequality) *Let $(\Omega, \mathcal{S}, \mu)$ be a measure space and $f, g : \Omega \to \mathbb{R}^+$ two density functions with respect to μ with positive function values. Let further the functions*

$$\eta_f : \Omega \to \mathbb{R}, \quad \omega \mapsto f(\omega) \, \mathrm{ld}(f(\omega))$$
$$\eta_{fg} : \Omega \to \mathbb{R}, \quad \omega \mapsto f(\omega) \, \mathrm{ld}(g(\omega))$$

(μ-)be quasi-integrable, then the following applies:

$$- \int \eta_f d\mu \leq - \int \eta_{fg} d\mu. \qquad \lhd$$

The required measurability properties of the functions η_f, η_{fg} are—as already mentioned—ensured by the continuity of the function ld and by the $\mathcal{S}$-$\mathcal{B}$-measurability of f and g. Let $x_0 \in (0, \infty)$ be, then the function

$$t_{x_0} : \mathbb{R} \to \mathbb{R}, \quad x \mapsto \frac{1}{x_0 \ln(2)} x + \mathrm{ld}(x_0) - \frac{1}{\ln(2)}$$

represents the tangent to the graph of ld at the point $(x_0, \mathrm{ld}(x_0))$. Since ld is a strictly concave function, for every $x_0 \in (0, \infty)$:

$$\mathrm{ld}(x) \leq t_{x_0}(x) \quad \text{for all} \quad x \in (0, \infty).$$

Now if we choose $x_0 = 1$, it follows:

$$\mathrm{ld}(x) \leq \frac{1}{\ln(2)} x - \frac{1}{\ln(2)} \quad \text{for all} \quad x \in (0, \infty).$$

Now let

$$x = \frac{p}{q} \quad \text{with} \quad p > 0, \ q > 0,$$

be, then the above inequality results in:

$$q - q \,\mathrm{ld}(q) \ln(2) \leq p - q \,\mathrm{ld}(p) \ln(2) \quad \text{with} \quad p > 0, \ q > 0.$$

If we set

$$q = f(\omega) \quad \text{and} \quad p = g(\omega),$$

then

$$f(\omega) - f(\omega) \,\mathrm{ld}(f(\omega)) \ln(2) \leq g(\omega) - f(\omega) \,\mathrm{ld}(g(\omega)) \ln(2) \quad \text{for all} \quad \omega \in \Omega.$$

applies. Now the right and left side of this inequality are integrated (μ-) integrated. As with the integration of elementary functions, the inequality remains and due to

$$\int f d\mu = \int g d\mu = 1$$

the assertion follows.   **q.e.d.**

Let the following be given: a measure space $(\Omega, \mathcal{S}, \mu)$, real numbers $a_1, \ldots, a_k$, $k \in \mathbb{N}$, and $\mathcal{S}$-$\mathcal{B}$-measurable functions

$$g_i : \Omega \to \mathbb{R}, \quad i = 1, \ldots, k,$$

with the following properties:

(1) The $\mathcal{S}$-$\mathcal{B}$-measurable function

$$h : \Omega \to \mathbb{R}, \quad \omega \mapsto \exp\left(\sum_{i=1}^{k} a_i g_i(\omega)\right)$$

is (μ-)integrable (and thus $0 < \int h d\mu < \infty$).

(2) With the density

$$d^* : \Omega \to \mathbb{R}^+, \quad \omega \mapsto \frac{h(\omega)}{\int h d\mu}$$

the following holds:

$$\int d^* g_i d\mu = b_i \in \mathbb{R}, \quad i = 1 \ldots, k,$$

we consider the set

$$\mathcal{D} := \big\{ d : \Omega \to \mathbb{R}^+; \quad d \text{ is a density with resp. to } \mu,$$
$$\eta_d : \Omega \to \mathbb{R}, \omega \mapsto d(\omega)\,\mathrm{ld}(d(\omega))$$
$$\text{and}$$
$$\eta_{dd^*} : \Omega \to \mathbb{R}, \quad \omega \mapsto d(\omega)\,\mathrm{ld}(d^*(\omega))$$
$$\text{are } (\mu\text{-})\text{quasi-integrable}\big\}$$

and the maximization problem

$$\max_{d \in \mathcal{D}} \left\{ -\int \eta_d d\mu; \int d g_1 d\mu = b_1 \right.$$

$$\vdots$$

$$\left. \int d g_k d\mu = b_k \right\}$$

The Gibbs inequality yields:

$$-\int \eta_d d\mu \le -\int \eta_{dd^*} d\mu$$

$$= -\int \left(-d\,\mathrm{ld}\left(\int h d\mu \right) \right) d\mu - \frac{1}{\ln(2)} \int \left(d \cdot \sum_{i=1}^{k} a_i g_i \right) d\mu$$

$$= \mathrm{ld}\left(\int h d\mu \right) - \frac{1}{\ln(2)} \sum_{i=1}^{k} a_i \int d g_i d\mu$$

$$= \mathrm{ld}\left(\int h d\mu \right) - \frac{1}{\ln(2)} \sum_{i=1}^{k} a_i b_i$$

$$= -\int \eta_{d^*} d\mu.$$

and thus the solution d^*.

Example 9.14 Using the measure space $([0, \infty), \mathcal{B}_{[0,\infty)}, \lambda_{[0,\infty)})$ with

$$\mathcal{B}_{[0,\infty)} := \{A \cap [0, \infty); \ A \in \mathcal{B}\} \quad \text{and} \quad \lambda_{[0,\infty)} : \mathcal{B}_{[0,\infty)} \to \mathbb{R}_0^+, A \mapsto \lambda(A)$$

we consider the function

$$g : [0, \infty) \to \mathbb{R}, \quad x \mapsto -x.$$

The function

$$h : \mathbb{R} \to \mathbb{R}, \quad x \mapsto \exp(ag(x))$$

is integrable for all $a > 0$ ($\lambda_{[0,\infty)}$-) and it holds:

$$\int (\exp \circ (ag)) d\lambda_{[0,\infty)} = \int\limits_0^\infty \exp(-ax)dx = \frac{1}{a}.$$

Thus,

$$d^* : \mathbb{R} \to \mathbb{R}^+, \quad x \mapsto a \exp(-ax)$$

is the density with maximum differential entropy among all densities from

$$\mathcal{D} := \Big\{ d : \mathbb{R} \to \mathbb{R}^+; \quad d \text{ is a density with resp. to } \lambda_{[0,\infty)},$$
$$\eta_d : \mathbb{R} \to \mathbb{R}, x \mapsto d(x) \operatorname{ld}(d(x))$$
$$\text{and}$$
$$\eta_{dd^*} : \mathbb{R} \to \mathbb{R}, x \mapsto d(x) \operatorname{ld}(d^*(x))$$
$$\text{are } (\lambda_{[0,\infty)}\text{-)quasi-integrable} \Big\}$$

under the constraint

$$\int fg d\lambda_{[0,\infty)} = \int d^* g d\lambda_{[0,\infty)} \left(= \int\limits_0^\infty -x d^*(x)dx = -\frac{1}{a} \right) \quad \text{for all } f \in \mathcal{D}.$$

The densities d^*, $a > 0$, represent the **exponential distributions**. ◁

Example 9.15 (Global Optimization) Let $n \in \mathbb{N}, \Omega = [-1, 1]^n$ and

$$\mathcal{B}_{[-1,1]^n}^n := \{A \cap [-1, 1]^n; \ A \in \mathcal{B}^n\},$$

then with

$$\lambda_{[-1,1]^n}^n : \mathcal{B}_{[-1,1]^n}^n \to \mathbb{R}_0^+, \quad M \mapsto \lambda^n(M)$$

the triple $([-1, 1]^n, \mathcal{B}_{[-1,1]^n}^n, \lambda_{[-1,1]^n}^n)$ is a measure space. Since with

$$g_n : [-1, 1]^n \to \mathbb{R}, \quad \mathbf{x} \mapsto \sum\limits_{i=1}^n (4x_i^2 - \cos(8x_i) + 1)$$

the function

$$h_n : [-1, 1]^n \to \mathbb{R}, \quad \mathbf{x} \mapsto \exp(-g_n(\mathbf{x}))$$

$(\lambda^n_{[-1,1]^n}-)$integrable,

$$d^*_n : [-1, 1]^n \to \mathbb{R}^+, \quad \mathbf{x} \mapsto \frac{\exp(-g_n(\mathbf{x}))}{\int (\exp \circ (-g_n)) d\lambda^n_{[-1,1]^n}}$$

is the density with maximum differential entropy among all densities from

$$\mathcal{D} := \left\{ d : [-1, 1]^n \to \mathbb{R}^+; \quad d \text{ is a density with resp. to } \lambda^n_{[-1,1]^n}, \right.$$

$$\eta_d : [-1, 1]^n \to \mathbb{R}, \mathbf{x} \mapsto d(\mathbf{x}) \operatorname{ld}(d(\mathbf{x}))$$
and
$$\eta_{dd^*} : [-1, 1]^n \to \mathbb{R}, \mathbf{x} \mapsto d(\mathbf{x}) \operatorname{ld}(d^*(\mathbf{x}))$$

$$\left. \text{are } (\lambda^n_{[-1,1]^n}-)\text{quasi-integrable} \right\}$$

under the constraint

$$\int dg_n d\lambda^n_{[-1,1]^n} = \int d^*_n g_n d\lambda^n_{[-1,1]^n} \quad \text{for all } d \in \mathcal{D}.$$

If one now wants to calculate the global minimum of g_n through local methods of nonlinear optimization, which generate a finite sequence of points with strictly monotonically decreasing function values (see for example [UlUl12]), ideal starting points for this are in the area $[-0.4, 0.4]^n$ (see Fig. 9.7 for $n = 1$).

Since information of this kind is generally not available a priori, one often chooses starting points from the domain of definition (here thus $[-1, 1]^n$) by pseudo-random numbers; in our example, one hits the said area with a probability $p = 0.4^n$. If one were to use pseudo-random numbers according to a distribution given by the density d^*_n (see for example d^*_1 in Fig. 9.8), the probability of hitting the desired area $[-0.4, 0.4]^n$ would be approximately given by

$$\hat{p} \approx 0.8^n (= 2^n \cdot 0.4^n).$$

The use of the distribution given by d^*_n thus increases the hit probability compared to pure random search in our example by the factor 2^n. The following Figs. 9.9 and 9.10 show g_2 and d^*_2 and illustrate this effect.

By using stochastic differential equations, it is now possible to simulate the maximization of differential entropy in the computer and thus make the density d^*_n usable for nonlinear optimization (see [Sch14]). This approach works for a large class of functions to be minimized.   ◁

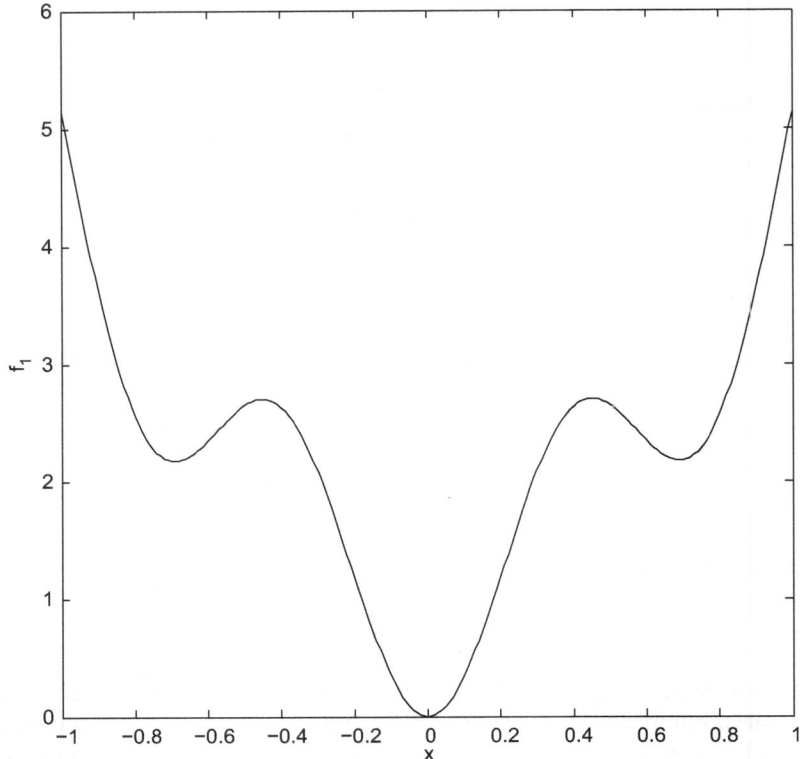

Fig. 9.7 Function f_1

9.4   Differential Entropy in Dynamic Systems

An important application of differential entropy is given by the analysis of dynamic systems. Thus, if (Ω, d) is a compact metric space and

$$T : \Omega \rightarrow \Omega$$

is a $\mathcal{B}_d$-$\mathcal{B}_d$-measurable mapping, where $\mathcal{B}_d$ represents the σ-algebra generated by the open subsets of Ω (open with respect to the metric d) (the index is now necessary because we will need the Borel σ-field over $\mathbb{R}$ denoted by $\mathcal{B}$), we obtain the dynamic system $(\Omega, \mathcal{B}_d, T)$. Furthermore, we assume the existence of a measure

$$\mu : \mathcal{B}_d \rightarrow [0, \infty)$$

such that

$$\mu(A) = 0 \quad \Longrightarrow \quad \mu(T^{-1}(A)) = 0 \quad \text{for all} \quad A \in \mathcal{B}_d.$$

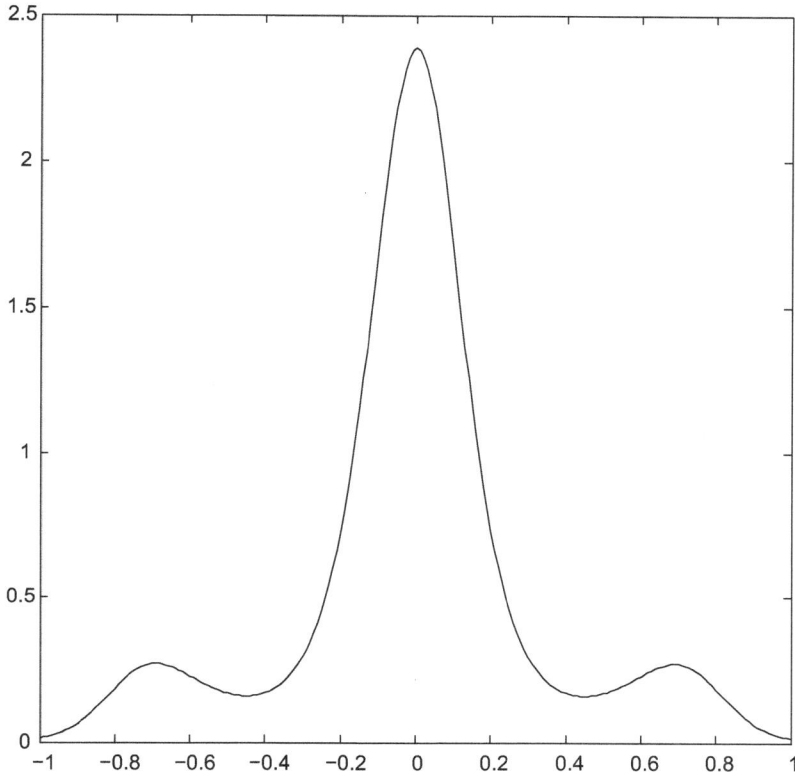

Fig. 9.8 Density d_1^*

If this condition is fulfilled, T is referred to as **nonsingular** with respect to μ (obviously, T is nonsingular with respect to any invariant measure).

The task now is to find a $\mathcal{B}_d$-$\bar{\mathcal{B}}$-measurable function

$$f : \Omega \to \mathbb{R}_0^+ \cup \{\infty\}$$

with the following properties:

(i)
$$\int f d\mu = 1.$$

(ii) The probability measure

$$\mathbb{P}_f : \mathcal{B}_d \to [0, 1], \quad A \mapsto \int_A f d\mu$$

is an invariant measure.

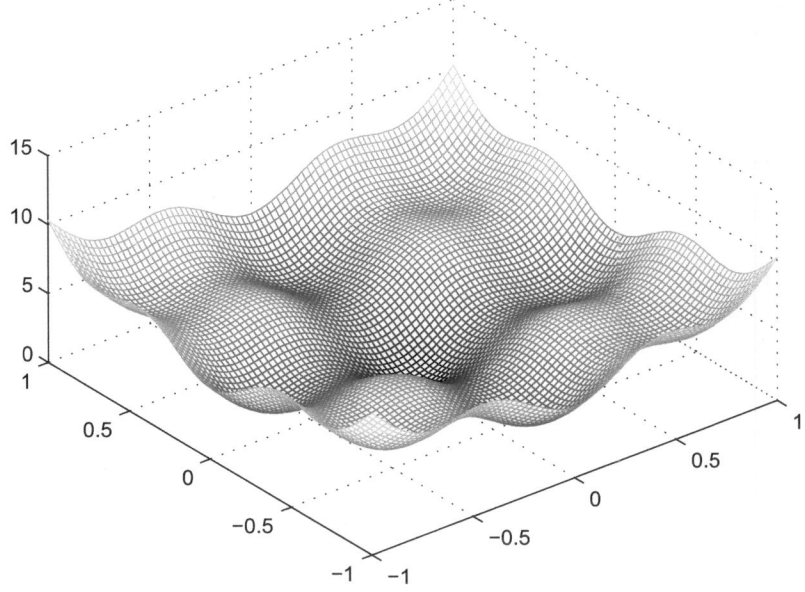

Fig. 9.9 Function g_2

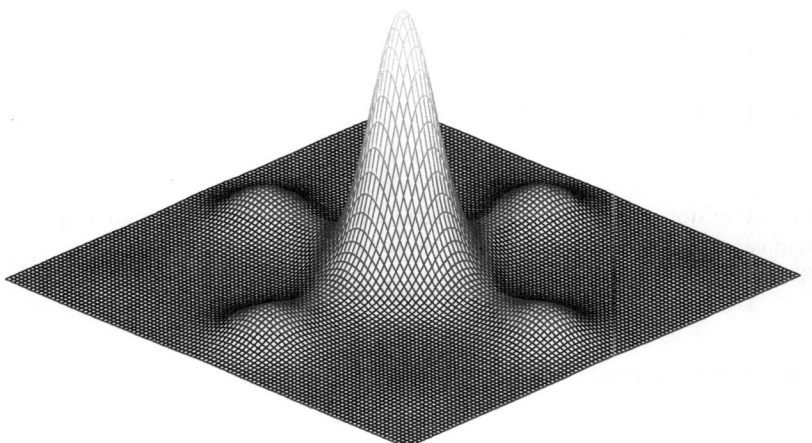

Fig. 9.10 Density d_2^*

So, we are looking for an invariant measure $\mathbb{P}_f$, which is given by a density f (however, with possible function values equal to ∞) with respect to μ. The differential entropy of f can then be used as an important characteristic size of the dynamic system.

To solve this problem, one considers the set

$$L_\mu^1 := \left\{ g : \Omega \to \mathbb{R}_0^+ \cup \{\infty\}; \ g \text{ is } \mathcal{B}_d\text{-}\bar{\mathcal{B}}\text{-measurable and } \int g \, d\mu = 1 \right\}.$$

First of all, it should be noted that with each function $h \in L_\mu^1$ through

$$\mathbb{P}^h : \mathcal{B}_d \to [0, 1], \quad A \mapsto \int_{T^{-1}(A)} h \, d\mu$$

a probability measure on $\mathcal{B}_d$ is given. Now the question arises about the existence of a function $\hat{h} \in L_\mu^1$ such that the probability measure $\mathbb{P}^h$ can be written in the form

$$\mathbb{P}^h : \mathcal{B}_d \to [0, 1], \quad A \mapsto \int_A \hat{h} \, d\mu = \int_{T^{-1}(A)} h \, d\mu$$

. Since T is non-singular with respect to μ, under a special condition on μ (the so-called σ-finiteness), an important theorem of measure and integration theory, the Radon-Nikodym theorem (see [Klenke05]), can be used, which guarantees the existence of $\hat{h}$; if there is a second function $h_1 \in L_\mu^1$ with

$$\mathbb{P}^h(A) = \int_A h_1 \, d\mu = \int_{T^{-1}(A)} h \, d\mu \quad \text{for all } A \in \mathcal{B}_d,$$

then it is also known from the Radon-Nikodym theorem that there is a set $N \in \mathcal{B}_d$ with $\mu(N) = 0$ and

$$\hat{h}(\omega) = h_1(\omega) \quad \text{for all} \quad \omega \in N^c \quad \text{(equality } \mu\text{-almost everywhere).}$$

If μ is σ-finite, i.e. there is a sequence $\{B_i\}_{i \in \mathbb{N}}$ of sets from $\mathcal{B}_d$ with

$$\mu(B_i) < \infty, \ i \in \mathbb{N}, \quad \text{and} \quad \bigcup_{i=1}^\infty B_i = \Omega,$$

then there exists a mapping

$$\mathfrak{F} : L_\mu^1 \to L_\mu^1 \quad \text{with} \quad \int_A \mathfrak{F}(h) \, d\mu = \int_{T^{-1}(A)} h \, d\mu \quad \text{for all} \quad A \in \mathcal{B}_d.$$

The mapping $\mathfrak{F}$ is referred to as the **Frobenius-Perron Operator** and is examined in more detail in [LasMac95] and [DingZhou09]. Since we are looking for an invariant measure

$$\mathbb{P}_f : \mathcal{B}_d \rightarrow [0, 1], \quad A \mapsto \int_A f d\mu$$

with $f \in L_\mu^1$, it is now obvious that we need to look for a fixed point

$$\mathfrak{F}(f) = f$$

of the Frobenius-Perron Operator for this purpose.

Example 9.16 (Logistic Transformation) Let's return to the dynamic system

$$([0, 1], \mathcal{B}_d, T)$$

with

$$T : [0, 1] \rightarrow [0, 1], \quad x \mapsto 4x(1 - x) \quad \text{(logistic transformation)},$$

where the metric, as usual with real numbers, is given by

$$d : [0, 1] \times [0, 1] \rightarrow \mathbb{R}, \quad (x, y) \mapsto |x - y|$$

, the Lebesgue measure $\lambda_{[0,1]}$ restricted to the interval $[0, 1]$ is obviously σ-finite. Now consider the Frobenius-Perron operator

$$\mathfrak{F} : L_{\lambda_{[0,1]}}^1 \rightarrow L_{\lambda_{[0,1]}}^1 \quad \text{with} \quad \int_A \mathfrak{F}(h) d\lambda_{[0,1]} = \int_{T^{-1}(A)} h d\lambda_{[0,1]} \quad \text{for all} \quad A \in \mathcal{B}_d,$$

because of

$$T^{-1}([0, x]) = \left[0, \frac{1}{2} - \frac{1}{2}\sqrt{1 - x}\right] \cup \left[\frac{1}{2} + \frac{1}{2}\sqrt{1 - x}, 1\right] \quad \text{for} \quad x \in [0, 1]$$

and with

$$f_0 : [0, 1] \rightarrow \mathbb{R}_0^+ \cup \{\infty\}, \quad \omega \mapsto 1 :$$

$$f_1 := \mathfrak{F}(f_0) : [0, 1] \rightarrow \mathbb{R}_0^+ \cup \{\infty\}, \quad \omega \mapsto \begin{cases} \frac{1}{2\sqrt{1-\omega}} & \text{if } 0 \leq \omega < 1 \\ \infty & \text{if } \omega = 1 \end{cases}$$

(see Fig. 9.11), while

$$f_2 := \mathfrak{F}(f_1) : [0, 1] \rightarrow \mathbb{R}_0^+ \cup \{\infty\},$$

$$\omega \mapsto \begin{cases} \frac{\sqrt{2}}{8\sqrt{1-\omega}} \left(\frac{1}{\sqrt{1+\sqrt{1-\omega}}} + \frac{1}{\sqrt{1-\sqrt{1-\omega}}} \right) & \text{if } 0 < \omega < 1 \\ \infty & \text{if } \omega = 0 \\ \infty & \text{if } \omega = 1 \end{cases}$$

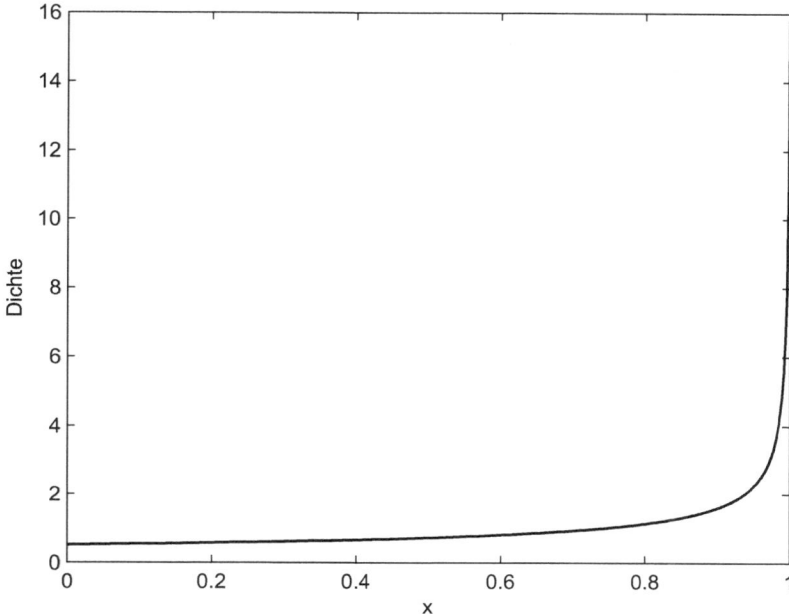

Fig. 9.11 Function f_1

(see Fig. 9.12).

In [UlNeu47], the sought invariant measure was calculated by

$$f = \mathfrak{F}(f) : [0, 1] \to \mathbb{R}_0^+ \cup \{\infty\}, \quad \omega \mapsto \begin{cases} \frac{1}{\pi \sqrt{\omega(1-\omega)}} & \text{if } 0 < \omega < 1 \\ \infty & \text{if } \omega = 0 \\ \infty & \text{if } \omega = 1 \end{cases}$$

(see Fig. 9.13). ◁

Now, motivated by example 9.4, we consider the fixed point iteration

$$f_{n+1} = \mathfrak{F}(f_n), \quad n \in \mathbb{N}_0,$$

for a dynamic system $(\Omega, \mathcal{B}_d, T)$ with σ-finite measure μ on $\mathcal{B}_d$, a non-singular transformation T with respect to μ, and the corresponding Frobenius-Perron operator $\mathfrak{F}$.

The following considerations serve to provide a small insight into the analysis of densities in dynamic systems and to develop a sense for the complexity of the question. A comprehensive treatment of this question would fill an entire book, without having addressed the deep functional-analytical prerequisites.

In order to be able to prove the existence and uniqueness of a fixed point f of the Frobenius-Perron operator and to be able to investigate the convergence of the

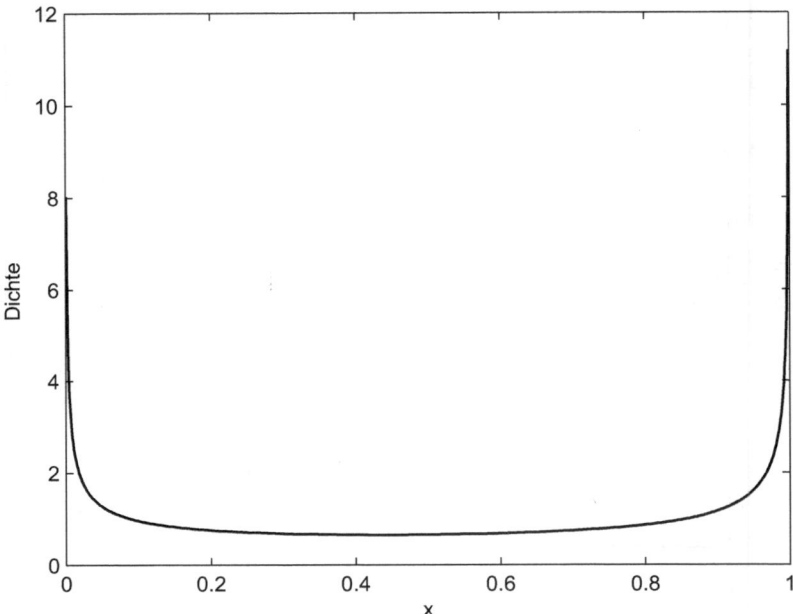

Fig. 9.12 Function f_2

sequence $\{f_n\}_{n\in\mathbb{N}_0}$, additional prerequisites for the dynamic system are needed; we only consider a classic scenario here:

Let $(\Omega, \mathcal{B}_d, T)$ be a dynamic system, μ a σ-finite measure on $\mathcal{B}_d$ and T non-singular with respect to μ; furthermore, let T be **ergodic** with respect to μ, i.e.

$$\mu(A) = 0 \quad \text{or} \quad \mu(A^c) = 0 \quad \text{for all} \quad A \in \mathcal{B}_d \text{ with } A = T^{-1}(A),$$

then there is at most one fixed point f of the Frobenius-Perron operator $\mathfrak{F}$ (see [LasMac95]). The condition of ergodicity means again that it is impossible to divide the dynamic system into two T-invariant sets $A_1, A_2 \in \mathcal{B}_d$ with

$$A_1 \cup A_2 = \Omega \quad \text{and} \quad \mu(A_1) > 0 \text{ and } \mu(A_2) > 0$$

into two dynamic systems

$$(A_1, \{M \cap A_1; \ M \in \mathcal{B}_d\}, T_{|A_1}) \quad \text{and} \quad (A_2, \{M \cap A_1; \ M \in \mathcal{B}_d\}, T_{|A_2})$$

and examine them separately.

To now be able to prove the existence of a fixed point f, one examines the **Cesàro means**

$$\frac{1}{k} \sum_{i=0}^{k-1} \mathfrak{F}^i(g), \quad k \in \mathbb{N}, \quad g \in L^1_\mu,$$

where

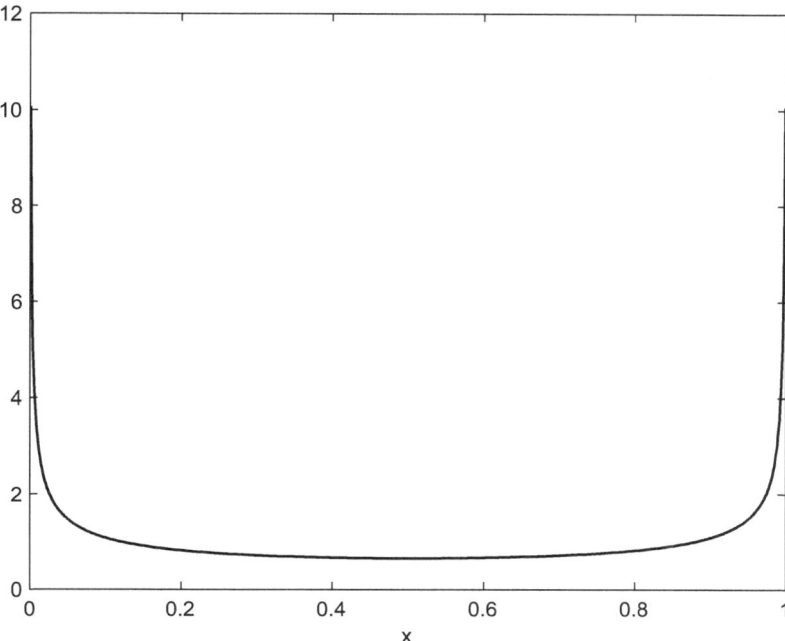

Fig. 9.13 Function f

$$\mathfrak{F}^i(g) := \mathfrak{F}(\mathfrak{F}^{i-1}(g)) \quad \text{and} \quad \mathfrak{F}^0(g) := g, \quad i \in \mathbb{N}, \quad g \in L_\mu^1.$$

According to the **Ergodic theorem of Kakutani-Yoshida** a fixed point f exists, if there is an $h \in L_\mu^1$ such that there is a weakly convergent subsequence in L_μ^1 for

$$\left\{ \frac{1}{k} \sum_{i=0}^{k-1} \mathfrak{F}^i(h) \right\}_{k \in \mathbb{N}}$$

(see again LasMac95). The limit is then given by the fixed point f. If this is the case, then:

$$\lim_{k \to \infty} \int \left| \frac{1}{k} \sum_{i=0}^{k-1} \mathfrak{F}^i(h) - f \right| d\mu = 0$$

(Theorem 5.1.1 in [DingZhou09]).

Since the logistic transformation

$$([0, 1], \mathcal{B}_d, T) \quad \text{mit} \quad \mu = \lambda_{[0,1]}$$

fulfills all the prerequisites just presented, the results documented in Example 9.4 are obvious.

Conditional Expectations

10.1 Existence and Uniqueness

In the investigation of sufficient statistics in Sect. 5.1 conditional probabilities

$$\mathbb{P}^B : \mathcal{P}(\Omega) \to [0,1], \ A \mapsto \frac{\mathbb{P}(A \cap B)}{\mathbb{P}(B)}$$

played an important role. The necessary prerequisite $\mathbb{P}(B) > 0$ was uncritical in the discrete probability spaces considered. In order to now investigate the question of sufficient statistics within the framework of general probability spaces, a generalization of the previously considered conditional probabilities is necessary. Starting from a probability space $(\Omega, \mathcal{S}, \mathbb{P})$ we consider a numerical, ($\mathbb{P}$-) integrable random variable

$$X : \Omega \to \bar{\mathbb{R}}.$$

The integral

$$\mathbb{E}(X) := \int X d\mathbb{P}$$

is referred to as the **expected value** of X. The mapping

$$Y : \Omega \to \bar{\mathbb{R}}, \quad \omega \mapsto \mathbb{E}(X)$$

is measurable for every σ-field $\mathcal{G}$ over Ω $\mathcal{G}$-$\bar{\mathcal{B}}$-measurable. Thus, $\{\emptyset, \Omega\}$ is the smallest of all σ-fields $\mathcal{G}$ over Ω, for which $Y\mathcal{G}$-$\bar{\mathcal{B}}$-measurable, and it holds:

$$\int_A Y d\mathbb{P} = \int_A X d\mathbb{P} \quad \text{for all } A \in \{\emptyset, \Omega\}.$$

© The Author(s), under exclusive license to Springer-Verlag GmbH, DE, part of Springer Nature 2024
S. Schäffler, *Mathematics of Information*, Mathematics Study Resources 9,
https://doi.org/10.1007/978-3-662-69102-1_10

If X is also $\{\emptyset, \Omega\}$-$\bar{B}$-measurable, then there exists a set $N \in S$ with $\mathbb{P}(N) = 0$ (a so-called a $\mathbb{P}$-null set) and it holds

$$X(\omega) = \mathbb{E}(X) = Y(\omega) \quad \text{for all } \omega \in N^c.$$

Another notation for this is:

$$X = \mathbb{E}(X) = Y \quad \mathbb{P}\text{-almost surely}$$

(analogously, we use this notation generally for statements that are true for all ω from the complement of a $\mathbb{P}$-null set). When transitioning from X to Y, the distribution of X is preserved. However, if X is not $\{\emptyset, \Omega\}$-$\bar{B}$-measurable, then when transitioning from X to Y, we have reduced the knowledge about the distribution of X to the expected value of X. We want to quantify this reduction as follows:

Is $\mathcal{F}$ the set of all σ-fields over Ω such that X is measurable for all $\mathcal{C} \in \mathcal{F}\mathcal{C}$-$\bar{B}$-measurable, then as is well known,

$$\sigma(X) := \bigcap_{\mathcal{C} \in \mathcal{F}} \mathcal{C}$$

is the smallest σ-field over Ω for which $X\mathcal{C}$-$\bar{B}$-measurable. Of course, $\{\emptyset, \Omega\} \subseteq \sigma(X)$ applies. If now $\{\emptyset, \Omega\} = \sigma(X)$, we have not lost any knowledge. The loss of knowledge when transitioning from X to Y (regarding the distribution of X) increases with the number of sets $A \in \sigma(X)$, $A \notin \{\emptyset, \Omega\}$. So we use σ-fields (in our example $\{\emptyset, \Omega\}$ and $\sigma(X)$) to characterize this loss. This approach reminds of the explanations at the end of section 7.2.

In the following, we consider the reverse path by prescribing a reduction of knowledge about the distribution of X in the form of a σ-field $\mathcal{C} \subseteq S$ over Ω and considering a $\mathcal{C}$-$\bar{B}$-measurable numerical random variable $Y : \Omega \to \bar{\mathbb{R}}$ for which the following applies:

$$\int_A Y d\mathbb{P} = \int_A X d\mathbb{P} \quad \text{for all } A \in \mathcal{C}.$$

Theorem and Definition 10.1 (Conditional Expectation) *Let* $(\Omega, S, \mathbb{P})$ *a probability space and X a numerical, $(\mathbb{P}\text{-})$integrable random variable, then for every σ-field $\mathcal{C} \subseteq S$ over Ω there exists a $\mathcal{C}$-$\bar{B}$-measurable numerical random variable*

$$Y : \Omega \to \bar{\mathbb{R}}$$

with:

$$\int_A Y d\mathbb{P} = \int_A X d\mathbb{P} \quad \text{for all } A \in \mathcal{C}.$$

For two $\mathcal{C}$-$\bar{B}$-measurable numerical random variables

$$Y, Y_1 : \Omega \to \bar{\mathbb{R}},$$

that satisfy the above equation,

$$Y = Y_1 \quad \mathbb{P}\text{-almost surely.}$$

applies. Any C-$\bar{\mathcal{B}}$-measurable numerical random variable

$$Z : \Omega \to \bar{\mathbb{R}},$$

for which $\mathbb{P}$-almost surely $Z = Y$ applies, is denoted by $\mathbb{E}(X|\mathcal{C})$ and is called the **conditional expectation** *of X under C. Therefore, the random variable $\mathbb{E}(X|\mathcal{C})$ is uniquely determined except for $\mathbb{P}$-almost sure equality. A fixed Z is called a version of $\mathbb{E}(X|\mathcal{C})$.*   ◁

The proof essentially consists of applying the already mentioned Radon-Nikodym theorem (see [Klenke05]).

Starting from a probability space $(\Omega, \mathcal{S}, \mathbb{P})$ and a measure space $(\Omega', \mathcal{S}')$ we now examine an $\mathbb{P}$-integrable numerical random variable

$$X : \Omega \to \bar{\mathbb{R}}$$

and $n \in \mathbb{N}$ random variables

$$Z_1, \ldots, Z_n : \Omega \to \Omega'.$$

With $\sigma(Z_1, \ldots, Z_n)$ we denote the smallest of all σ-fields $\mathcal{C}$ over Ω, for which the random variables $Z_1, \ldots, Z_n$ $\mathcal{C}$-$\mathcal{S}'$-measurable are. If one is now interested in the conditional expectation $\mathbb{E}(X|\sigma(Z_1, \ldots, Z_n))$, one writes for it $\mathbb{E}(X|Z_1, \ldots, Z_n)$.

Under the conditions of sentence and definition 10.1, the following properties of conditional expectations arise:

(i) $\mathbb{E}(\mathbb{E}(X|\mathcal{C})) = \mathbb{E}(X)$.
(ii) If $X\mathcal{C}$-$\bar{\mathcal{B}}$-measurable, then $\mathbb{E}(X|\mathcal{C}) = X$ $\mathbb{P}$-almost surely follows.
(iii) If $X(\omega) = \alpha \in \mathbb{R}$ for all $\omega \in \Omega$, then $\mathbb{E}(X|\mathcal{C}) = \alpha$ $\mathbb{P}$-almost surely applies.

If now $Z : \Omega \to \bar{\mathbb{R}}$ is another $\mathbb{P}$-integrable numerical random variable, we obtain:

(1)For all $\alpha, \beta \in \mathbb{R}$ with

$$\alpha X + \beta Z : \Omega \to \bar{\mathbb{R}} \quad \text{(i.e., not } '\infty - \infty')$$

applies:

$$\mathbb{E}(\alpha X + \beta Z|\mathcal{C}) = \alpha \mathbb{E}(X|\mathcal{C}) + \beta \mathbb{E}(Z|\mathcal{C}) \quad \mathbb{P}\text{-almost surely,}$$

(2)Now let $X \leq Z\mathbb{P}$-almost surely, then applies:

$$\mathbb{E}(X|\mathcal{C}) \leq \mathbb{E}(Z|\mathcal{C}) \quad \mathbb{P}\text{-almost surely.}$$

The class of conditional expectations defined in the following definition is particularly important.

Definition 10.2 (conditional probability) *Let $(\Omega, \mathcal{S}, \mathbb{P})$ be a probability space and A an arbitrary set from the σ-field $\mathcal{S}$, then for each σ-field $\mathcal{C} \subseteq \mathcal{S}$ the conditional expectation $\mathbb{E}(I_A | \mathcal{C})$ is referred to as conditional probability and is represented in the form $\mathbb{P}(A|\mathcal{C})$, where it is known that*

$$I_A : \Omega \to \mathbb{R}, \quad \omega \mapsto \begin{cases} 1 & \text{if } \omega \in A \\ 0 & \text{else} \end{cases}.$$

$\triangleleft$

We will establish a connection between probability measures

$$\mathbb{P}^B : \mathcal{S} \to [0,1], \; A \mapsto \frac{\mathbb{P}(A \cap B)}{\mathbb{P}(B)}, \quad \mathbb{P}(B) > 0, B \in \mathcal{S}$$

and special conditional probabilities in the sense of Definition 10.1 in the following. This connection is the justification for the chosen terms "conditional expectation" and "conditional probability".

Factorization Lemma Let $(\Omega', \mathcal{S}')$ be a measurable space, Ω a non-empty set, $Y : \Omega \to \Omega'$ a mapping and $Z : \Omega \to \bar{\mathbb{R}}$ a numerical function. We denote by $\sigma(Y)$ the smallest of all σ-fields $\mathcal{C}$ over Ω, for which Y $\mathcal{C}$-$\mathcal{S}'$-measurable, then Z is exactly $\sigma(Y)$-$\bar{\mathcal{B}}$-measurable, if there is a $\mathcal{S}'$-$\bar{\mathcal{B}}$-measurable numerical function $g : \Omega' \to \bar{\mathbb{R}}$ with $Z = g \circ Y$.

Since Y $\mathcal{P}(\Omega)$-$\bar{\mathcal{B}}$-measurable, the existence of $\sigma(Y)$ is assured. If $Z = g \circ Y$, then Z as a concatenation of a $\sigma(Y)$-$\mathcal{S}'$-measurable and a $\mathcal{S}'$-$\bar{\mathcal{B}}$-measurable mapping $\sigma(Y)$-$\bar{\mathcal{B}}$-measurable. We divide the proof for the reversal into 3 parts:

(1) Let $Z = \sum\limits_{i=1}^{n} \alpha_i I_{A_i}$ with $\alpha_i \in \mathbb{R}_0^+$ and $A_i \in \sigma(Y)$ for all $i = 1, \dots, n$, then for every set A_i there is a set $A_i' \in \mathcal{S}'$ with $Y^{-1}(A_i') = A_i$, $i = 1, \dots, n$. Thus, $g = \sum\limits_{i=1}^{n} \alpha_i I_{A_i'}$ accomplishes the required.

(2) Let $Z(\omega) \geq 0$ be for all $\omega \in \Omega$, then there is a monotonically increasing sequence $\{e_k\}_{k \in \mathbb{N}}$ of non-negative elementary functions $e_k : \Omega \to \mathbb{R}_0^+$ with $e_k \uparrow Z$. According to part (1), for each function e_k, $k \in \mathbb{N}$, there is a non-negative, elementary function $g_k : \Omega' \to \mathbb{R}_0^+$ with $e_k = g_k \circ Y$. Since $Z = \sup\{e_k\}$, $g := \sup\limits_{k \in \mathbb{N}} \{g_k\}$ fulfills the requirement.

(3) For general Z we consider Z^+ and Z^-. According to part (2), there are two non-negative numerical functions $g', g'' : \Omega' \to \bar{\mathbb{R}}_0^+$ with

$$Z^+ = g' \circ Y \quad \text{and} \quad Z^- = g'' \circ Y.$$

Now we cannot switch to the difference of g' and g'' as this difference is not defined on the set

$$U' := \{\omega' \in \Omega'; \; g'(\omega') = \infty \wedge g''(\omega') = \infty\}.$$

But since

$$Z(\omega) = Z^+(\omega) - Z^-(\omega) = g'(Y(\omega)) - g''(Y(\omega)) \quad \text{for all} \quad \omega \in \Omega,$$

is $Y(\Omega) \cap U' = \emptyset$,

$$g : \Omega' \to \bar{\mathbb{R}}, \quad \omega' \mapsto \begin{cases} g'(\omega') - g''(\omega') & \text{for } \omega' \in U'^c \\ 0 & \text{else} \end{cases}$$

fulfills the requirement.

Now we consider our probability space $(\Omega, \mathcal{S}, \mathbb{P})$, a ($\mathbb{P}$-)integrable numerical random variable $X : \Omega \to \bar{\mathbb{R}}$, a measure space $(\Omega', \mathcal{S}')$ and a random variable $Y : \Omega \to \Omega'$. With

$$\mathbb{E}(X|Y) : \Omega \to \mathbb{R}$$

we denote a real version of the conditional expectation of X under $\sigma(Y)$. Since the real random variable $\mathbb{E}(X|Y) \sigma(Y)$-$\mathcal{B}$-measurable, there is, according to the factorization lemma (Lemma 1), a—depending on the chosen version $\mathbb{E}(X|Y)$—$\mathcal{S}'$-$\mathcal{B}$-measurable real function $g : \Omega' \to \mathbb{R}$ with

$$\mathbb{E}(X|Y) = g \circ Y.$$

This function g is uniquely determined up to $\mathbb{P}_Y$-null sets. We want to assume in the following that there is a $\hat{\omega}' \in \Omega'$ with $\{\hat{\omega}'\} \in \mathcal{S}'$ and

$$\mathbb{P}(\{\omega \in \Omega; \ Y(\omega) = \hat{\omega}'\}) > 0.$$

With

$$B := \{\omega \in \Omega; \ Y(\omega) = \hat{\omega}'\}$$

we obtain

$$\int_B g \circ Y d\mathbb{P} = \int g(\hat{\omega}') \cdot I_B d\mathbb{P} = g(\hat{\omega}') \cdot \mathbb{P}(B)$$

and

$$\int_B g \circ Y d\mathbb{P} = \int_B \mathbb{E}(X|Y) d\mathbb{P} = \int_B X d\mathbb{P}.$$

Thus, it follows:

$$\mathbb{E}(X|Y)(\omega) = g(\hat{\omega}') = \frac{1}{\mathbb{P}(B)} \int_B X d\mathbb{P} = \int X d\mathbb{P}^B =: \mathbb{E}(X|Y = \hat{\omega}') \text{ for all } \omega \in B.$$

If the random variable X has the special form $X = I_A$, $A \in \mathcal{S}$, then it follows:

$$\mathbb{E}(I_A|Y)(\omega) = \mathbb{P}(A|Y)(\omega) = \int I_A d\mathbb{P}^B = \mathbb{P}^B(A) =: \mathbb{P}(A|Y = \hat{\omega}') \text{ for all } \omega \in B.$$

The terms "conditional expectation" and "conditional probability" are justified by:

$$\mathbb{E}(X|Y)(\omega) = \int X d\mathbb{P}^B \quad \text{and} \quad \mathbb{P}(A|Y)(\omega) = \mathbb{P}^B(A) \quad \text{for all } \omega \in B$$

and for all real versions $\mathbb{E}(X|Y)$.

Definition 10.3 (conditional expectation under $Y = y$) Let $(\Omega, \mathcal{S}, \mathbb{P})$ be a probability space, $(\Omega', \mathcal{S}')$ a measurable space,

$$X : \Omega \to \bar{\mathbb{R}}$$

a ($\mathbb{P}$-)integrable numerical random variable, $Y : \Omega \to \Omega'$ a random variable and $g \circ Y$ a real version of $\mathbb{E}(X|Y)$ with $g : \Omega' \to \mathbb{R}$, then with $y := Y(\hat{\omega})$ for $\hat{\omega} \in \Omega$ the real number

$$\mathbb{E}(X|Y = y) := g(y) = (g \circ Y)(\hat{\omega})$$

is called the conditional expectation of X under $Y = y$.   ◁

Now that we have a suitable generalization of conditional probabilities for indicator functions, we can treat the question of sufficient statistics more generally in the following section. The specific calculation of a version of a conditional expectation always depends on the specific case considered. There is no universally applicable method.

10.2   Sufficiency

In Sect. 5.1 we started from the following initial situation:

Given is a triple $(\Omega, \mathcal{P}(\Omega), \mathbb{P}_\theta)$ consisting of a non-empty countable base set Ω, a non-empty set Θ, and a set

$$\{\mathbb{P}_\theta; \theta \in \Theta\}$$

of probability measures on $\mathcal{P}(\Omega)$; furthermore, an observed result $\hat{\omega} \in \Omega$ of the random experiment with $\mathbb{P}_\theta(\{\hat{\omega}\}) > 0$ for all $\theta \in \Theta$ is given. The fact that the conditional probabilities

$$\mathbb{P}_\theta^{\{\hat{\omega}\}} : \mathcal{P}(\Omega) \to [0, 1], \quad A \mapsto \frac{\mathbb{P}_\theta(\{\hat{\omega}\} \cap A)}{\mathbb{P}_\theta(\{\hat{\omega}\})} = \begin{cases} 1 & \text{if } \hat{\omega} \in A \\ 0 & \text{if } \hat{\omega} \notin A \end{cases}, \quad \theta \in \Theta$$

do not depend on θ, we interpreted as the knowledge of $\hat{\omega}$ is sufficient to make a decision about $\theta \in \Theta$. This led to the definition of a sufficient statistic

$$T : \Omega \to \tilde{\Omega}$$

for θ: For all

$$F = \{\omega \in \Omega; \ T(\omega) = \tilde{\omega}\}, \quad \tilde{\omega} \in \tilde{\Omega},$$

applies:

(1) $\mathbb{P}_\theta(F) > 0$ for all $\theta \in \Theta$,
(2) $\mathbb{P}_\theta^F(A)$ depends for all $A \in \mathcal{P}(\Omega)$ not on $\theta \in \Theta$.

Since for a set $(\Omega, \mathcal{S}, \mathbb{P}_\theta)$, $\theta \in \Theta$, of no longer necessarily discrete probability spaces the requirement $\mathbb{P}(F) > 0$ for the conditional probability measure $\mathbb{P}^F$ is problematic, we are now dependent on the following obvious modification.

Definition 10.4 (sufficient statistic, general case) Let Θ be a non-empty set and $(\Omega, \mathcal{S}, \mathbb{P}_\theta)$ for each $\theta \in \Theta$ a probability space. Furthermore, let $(\Omega', \mathcal{S}')$ be a measurable space and

$$T : \Omega \to \Omega'$$

a $\mathcal{S}\text{-}\mathcal{S}'$-measurable mapping, then T is **called a sufficient statistic** for θ, if for each event $A \in \mathcal{S}$ there is a version

$$\mathbb{P}_\bullet(A|T)$$

of the conditional probability $\mathbb{P}_\theta(A|T) = \mathbb{E}_\theta(I_A|T)$ that is independent of $\theta \in \Theta$. ◁
     We already know that

$$\mathbb{P}_\theta(A|T)(\omega) = \mathbb{P}_\theta^{\{\omega \in \Omega; \; T(\omega) = \hat{\omega}'\}}(A),$$

if

$$\mathbb{P}_\theta(\{\omega \in \Omega; \; T(\omega) = \hat{\omega}'\}) > 0.$$

Therefore, Definition 10.2 is compatible with Definition 5.1.
     Now let's consider a scenario that is relevant for applications. Based on the probability spaces

$$(\mathbb{R}^n, \mathcal{B}^n, \mathbb{P}_\theta), \quad \theta \in \Theta, n \in \mathbb{N}$$

we assume that the probability measures $\mathbb{P}_\theta$ for each $\theta \in \Theta$ are given by a density

$$f_\theta : \mathbb{R}^n \to \mathbb{R}_0^+$$

with respect to λ^n. Now let's consider a random variable

$$T : \mathbb{R}^n \to \mathbb{R},$$

it is known from mathematical statistics (see for example [CasBer01]) that T is sufficient for θ if the densities f_θ can be represented as follows:

$$f_\theta : \mathbb{R}^n \to \mathbb{R}_0^+, \quad \mathbf{x} \mapsto g_\theta(T(\mathbf{x}))h(\mathbf{x}), \quad \theta \in \Theta,$$

with corresponding functions $g_\theta : \mathbb{R} \to \mathbb{R}$ and $h : \mathbb{R}^n \to \mathbb{R}$.

Example 10.6 (Normal Distribution) Let $(\mathbb{R}^n, \mathcal{B}^n, \mathbb{P}_\theta)$, $n \in \mathbb{N}$ be probability spaces,

$$\theta \in \mathbb{R} =: \Theta,$$

and let the probability measures $\mathbb{P}_\theta$ for each $\theta \in \mathbb{R}$ be given by the density

$$f_\theta : \mathbb{R}^n \to \mathbb{R}^+, \quad (x_1, \ldots, x_n) \mapsto \prod_{i=1}^{n} \frac{e^{-\frac{(x_i-\theta)^2}{2}}}{\sqrt{2\pi}}$$

with respect to λ^n, then

$$T : \mathbb{R}^n \to \mathbb{R}, \quad (x_1, \ldots, x_n) \mapsto \frac{1}{n}\sum_{i=1}^{n} x_i$$

is a sufficient statistic for θ, since for all $\theta \in \mathbb{R}$ and $\mathbf{x} \in \mathbb{R}^n$ the following holds:

$$\prod_{i=1}^{n} \frac{e^{-\frac{(x_i-\theta)^2}{2}}}{\sqrt{2\pi}} = \underbrace{\frac{1}{(\sqrt{2\pi})^n} \exp\left(-\frac{n(T(\mathbf{x})-\theta)^2}{2}\right)}_{=:g_\theta(T(\mathbf{x}))} \cdot \underbrace{\exp\left(-\frac{\sum\limits_{i=1}^{n}(x_i - T(\mathbf{x}))^2}{2}\right)}_{h(\mathbf{x})}.$$

Therefore, if instead of $\mathbf{x} \in \mathbb{R}^n$ only $T(\mathbf{x}) \in \mathbb{R}$ is stored, no information about $\theta \in \mathbb{R}$ is lost. From stochastic theory (e.g., Klenke05), it is known that the image measures $\mathbb{P}_{\theta,T}$ of the random variable T are given by the densities

$$f_{\theta,T} : \mathbb{R} \to \mathbb{R}^+, \quad x \mapsto \frac{e^{-\frac{n(x-\theta)^2}{2}}}{\sqrt{\frac{2\pi}{n}}}.$$

Now, if we compare the differential entropies of the densities f_θ and $f_{\theta,T}$, the result is:

$$-\int f_\theta \operatorname{ld}(f_\theta)d\lambda^n = n\frac{\operatorname{ld}(2\pi e)}{2},$$

$$-\int f_{\theta,T} \operatorname{ld}(f_{\theta,T})d\lambda = \frac{\operatorname{ld}\left(\frac{2\pi e}{n}\right)}{2}.$$

The comparison of these two numbers also shows how much information can be saved when transitioning from $\mathbf{x} \in \mathbb{R}^n$ to $T(\mathbf{x}) \in \mathbb{R}$ without losing information about $\theta \in \mathbb{R}$. For $n = 100$, for example, the following applies:

$$-\int f_\theta \operatorname{ld}(f_\theta)d\lambda^{100} \approx 100 \cdot 2.047 = 204.7 > -\int f_{\theta,T} \operatorname{ld}(f_{\theta,T})d\lambda \approx -1.27.$$

Fig. 10.1 and 10.2 show the densities f_0 and $f_{0,T}$ (i.e., $\theta = 0$) for $n = 2$.   ◁

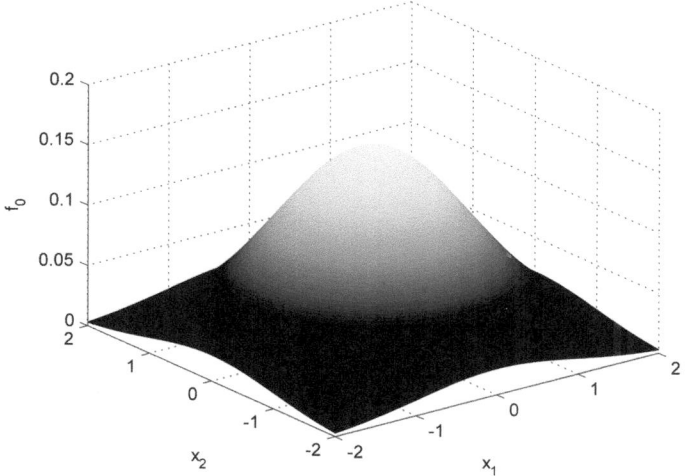

Fig. 10.1 Density f_0, $n=2$

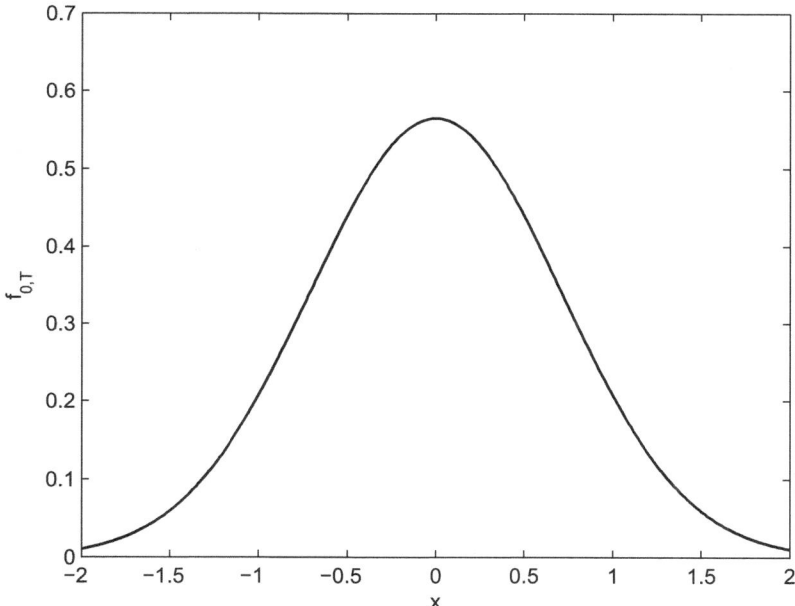

Fig. 10.2 Density $f_{0,T}$, $n=2$

References

[Ash65]   Ash, R.B.: Information Theory. Dover, New York (1965).
[Bau92]   Bauer, H.: Maß- und Integrationstheorie. de Gruyter, Berlin New York (1992).
[Bill65]   Billingsley, P.: Ergodic Theory and Information. John Wiley & Sons, New York, London, Sidney (1965).
[Buch10]   Buchmann, J.: Einführung in die Kryptographie. Springer, Berlin Heidelberg New York (2010).
[CasBer01]   Casella, G.; Berger, R.L.: Statistical Inference. Duxbury Press, Pacific Grove, CA (2001).
[CovTho91]   Cover, T.M.; Thomas, J.A.: Elements of Information Theory. Wiley & Sons, New York, Chichester, Brisbane, Toronto, Singapore (1991).
[Denker05]   Denker, M.: Einführung in die Analysis dynamischer Systeme. Springer, Berlin Heidelberg New York (2005).
[DingZhou09]   Ding, J.; Zhou A.: Statistical Properties of Deterministic Systems. Springer, Berlin Heidelberg New York (2009).
[Down11]   Downarowicz, T.: Entropy in Dynamical Systems. Cambridge University Press (2011).
[EinSch14]   Einsiedler, M.; Schmidt, K.: Dynamische Systeme. Birkhäuser, Basel (2014).
[Fi25]   Fisher, R.A.: Theory of statistical estimation. Proc. Camb. Phil. Soc., Vol. 22 (1925), pp. 700–725.
[For13]   Forster, O.: Algorithmische Zahlentheorie. Springer Fachmedien Wiesbaden (2013).
[Frie96]   Friedrichs, B.: Kanalcodierung. Springer, Berlin Heidelberg New York (1996).
[Gauß94]   Gauß, E.: WALSH-Funktionen. Teubner, Stuttgart (1994).
[HeiQua95]   Heise, W; Quattrocchi, P.: Informations- und Codierungstheorie. Springer, Berlin Heidelberg New York (1995).
[Held08]   Held, L.: Methoden der statistischen Inferenz. Spektrum Akademischer Verlag, Heidelberg (2008).
[HeHo74]   Henze, E.; Homuth, H.H.: Einführung in die Informationstheorie. Vieweg, Braunschweig (1974).
[Joh04]   Johnson, O.: Information Theory and The Central Limit Theorem. Imperial College Press, London (2004).
[Klenke05]   Klenke, A.: Wahrscheinlichkeitstheorie. Springer, Berlin Heidelberg New York (2005).
[Kom]   Komar, E.: Rechnergrundlagen. Skriptum FH Darmstadt.
[Kull97]   Kullback, S.: Information Theory and Statistics. Dover, New York (1997).
[LasMac95]   Lasota, A.; Mackey M.C.: Chaos, Fractals, and Noise: Stochastic Aspects of Dynamics. Springer, Berlin Heidelberg New York (1995).

[NieChu00]   Nielsen, M.A.; Chuang, I.L.: Quantum Computation and Quantum Information. Cambridge University Press (2000).

[OhmLü10]   Ohm, J.; Lüke, H.D.: Signalverarbeitung. Springer, Berlin Heidelberg New York (2010).

[Pas38]   Pascal, B.: Die Kunst zu überzeugen. Lambert Schneider, Berlin (1938).

[PötSob80]   Pötschke, D.; Sobik, F.: Mathematische Informationstheorie. Akademie-Verlag, Berlin (1980).

[Sch14]   Schäffler, S.: Globale Optimierung. Springer, Berlin Heidelberg New York (2014).

[ShWe63]   Shannon, C.E.; Weaver, W.: The Mathematical Theory of Communication. University of Illinois Press, Urbana and Chicago (1963).

[Stier10]   Stierstadt, K.: Thermodynamik. Springer, Berlin Heidelberg New York (2010).

[StSch09]   Sturm, T.F.; Schulze, J.: Quantum Computation aus algorithmischer Sicht. Oldenbourg, München (2009).

[UlNeu47]   Ulam, S.; von Neumann, J.: On combination of stochastic and deterministic processes. Bull. Amer. Math. Soc., 53 (1947), p. 1120.

[UlUl12]   Ulbrich, M.; Ulbrich, S.: Nichtlineare Optimierung. Birkhäuser, Basel (2012).

[Wagon85]   Wagon, S.: The Banach-Tarski Paradoxon. Cambridge University Press (1985).

[Wie61]   Wiener, N.: Cybernetics or Control and Communication in the Animal and the Machine. The MIT Press, Cambridge Massachusetts, second edition (1961).